DEVELOPMENT AND EVOLUTION

DEVELOPMENT

AND

EVOLUTION

INCLUDING

PSYCHOPHYSICAL EVOLUTION, EVOLUTION BY
ORTHOPLASY, AND THE THEORY OF
GENETIC MODES

BY

JAMES MARK BALDWIN

Ph.D. Princeton, Hon. D.Sc. Oxon., LL.D. Glasgow
STUART PROFESSOR IN PRINCETON UNIVERSITY

THE BLACKBURN PRESS

Development and Evolution
Including Psychophysical Evolution, Evolution by Orthoplasy,
and the Theory of Genetic Modes

ISBN-10: 1-930665-13-X
ISBN-13: 978-1-930665-13-2

Library of Congress Control Number: 2004106378

THE BLACKBURN PRESS
P. O. Box 287
Caldwell, New Jersey   07006 U.S.A.
973-228-7077
www.BlackburnPress.com

To My Friends

C. LLOYD MORGAN, H. F. OSBORN, E. B. POULTON

PREFACE

The present volume fulfils in a general way the intention, expressed in the preface to the first edition[1] of the work *Social and Ethical Interpretations*, of taking up some of the biological problems most closely connected with psychological ones and falling under the general scope of the genetic method. General biology is to-day mainly theory of evolution, and its handmaid is theory of individual development.

The composition of the work — like that of the companion volumes — has been gradual, and the positions taken have been in many cases already presented in journals under various dates since 1895. This is especially true of the matter contained in Part II., regarding which a word of more detailed explanation is necessary.

Since the first publication of the position, called in these pages and earlier 'Organic Selection,' by three writers independently, — Professor H. F. Osborn, Principal Lloyd Morgan, and myself, — considerable discussion has arisen about the theory, its meaning and value, and the original papers announcing the point of view have been under somewhat close inspection. The demand for reprints of these papers, in my own case — to speak only of my own case — has made it seem advisable to have them put in some available form much as they originally appeared. Despite the difficulties in the way of doing this, arising

---

[1] Reprinted in the third edition (1902).

mainly from the lack of continuity and the overlapping which such papers would present when printed together under one cover, I have still determined upon this course. It was my first intention to write a general introduction to evolution, — an exposition and criticism of the great theories, — and indeed such an intention is embodied in a contract with the publishers of the 'Science Series'; but it now becomes necessary to make that undertaking a separate affair, since this volume makes no pretence to completeness from such a point of view. I may add that that purpose is indeed, to my mind, excellently served by Professor H. W. Conn's able and readable book, *The Method of Evolution*, along with which students may take up also with profit the work *Problems of Evolution*, by F. W. Headley.

This change of plan once determined upon, it seemed highly desirable that the original papers of Professors Osborn and Lloyd Morgan should be liberally drawn upon, both in the interest of coöperation — from the first most cordial and friendly — and in that of advantage to our common views; for their positions were reached from quite different lines of approach, and the theory gains very much from this diversity of presentation. I accordingly secured their consent to my making liberal quotations from their papers; and, as I proceeded, it occurred to me that instead of making detached citations here and there the reader would profit more by longer quotations, — and, indeed, that the authorities quoted would thus be much more adequately presented. Hence the full citations from these authorities in Appendix A.[1] This method once

---

[1] The fulness of my personal recognition of them, as well as of another, is expressed, though still inadequately, in the dedication of this volume. At the same time, these writers are, of course, in no way implicated in the views of the book, except as their own statements are quoted.

approved, it became consonant with it to include the additional quotations from Professor Poulton (Appendix A)
and Professors Conn and Headley (Appendix B), —
all of which serve as substitutes for frequent separate
citations in various parts of the text, but gain in force
by this 'solid' form of presentation. I am under obliga-
tions to all these writers (and also to their publishers) for
their generous permission to make such free use of their
writings. Principal Lloyd Morgan has also favoured me
with the concise 'new statement' — as I call it for convenience of reference — of his views, printed, with the
citations mentioned, in Appendix A.

The work thus becomes, so far as this portion of it is
concerned, a sort of handbook of the theory of 'Orthoplasy,'[1] — exhibiting its original forms of presentation and
reflecting its progress up to date. The defects of the
method, from the point of view of the 'continuous' reader,
are so evident that I hope the critic may not find it in his
heart, after these explanations, to 'rub it in.' The principal and obvious disadvantage is seen in certain necessary
repetitions. Yet these are always in the course of the
discussions of different phases of the larger topics; and
to the psychologist, at least, repetition has its pedagogical
justification. All readers are not equally mature; and
even to the least immature the saying 'here a little and
there a little' is still the formula of least exertion.

On the other hand, the remaining portions of the book,
Parts II. and III., are mostly new matter. Of this new
matter the things which are submitted by the writer with
solicitude — defined as 'hope with sufficient fear' — are
the exposition of 'Psychophysical Evolution' and the outline sketch of the 'Theory of Genetic Modes.' These are
more properly within the range of a professed psycholo-

---

[1] The theory of evolution which makes essential use of 'organic selection.'

gist's interests than are points in biology, and I am accordingly the less disinclined to cast them upon the water expecting some return after many days.

The relation of this volume to the two earlier ones is spoken of above. The close connection of the three volumes, all of which might have been made parts of a single larger work, renders necessary the repeated citation of each one of them in the others, in a way which may seem — and has seemed, to one critic — to be a case of a writer's liking 'to quote himself.' It is really, however, a matter of division of material — with separate publication of the parts — and the references are such as one usually finds from chapter to chapter in the course of one work. The interconnection of the topics it is, therefore, with the need of expounding them, for the sake of comprehensiveness, in this interconnection, that gives a somewhat personal look to these references.

J. M. B.

PRINCETON UNIVERSITY,
May, 1902.

CONTENTS

PART I

THE PROBLEM OF GENESIS

CHAPTER I

PSYCHOPHYSICAL EVOLUTION

CHAPTER II

COMPARATIVE CONCEPTIONS

CHAPTER III

THE DIRECTION OF EVOLUTION

Contents

PART II

THE METHOD OF EVOLUTION

CHAPTER IV

THE PLACE OF CONSCIOUSNESS IN EVOLUTION

CHAPTER V

HEREDITY AND INSTINCT (I.)

CHAPTER VI

HEREDITY AND INSTINCT (II.)

CHAPTER VII

PHYSICAL HEREDITY AND SOCIAL TRANSMISSION

CHAPTER VIII

A FACTOR IN EVOLUTION: ORGANIC SELECTION

Contents

Contents

CHAPTER XVIII

THE ORIGIN OF A 'THING' AND ITS NATURE

CHAPTER XIX

THE THEORY OF GENETIC MODES

Contents

DEVELOPMENT AND EVOLUTION

DEVELOPMENT AND EVOLUTION

PART I

THE PROBLEM OF GENESIS

CHAPTER I

PSYCHOPHYSICAL EVOLUTION

§ I. *Scope and Method*

THE point of view from which the questions taken up in the following pages are considered is still exclusively that of the earlier volumes of this series,[1] the genetic. But the broadening out of the range of discussion to include biological questions as well as psychological, makes our method now *Biogenetic* rather than *Psychogenetic* — a distinction made out in the volume on *Social and Ethical Interpretations.* It is not now, in these discussions, a question of the application of results, drawn from the mental life exclusively, to the larger problem of racial and social evolution; it is rather the interpretation of the whole series of facts drawn from all these spheres, examined with view to a general conception of genesis (subject to the self-imposed limitations indicated in the Preface).

The emphasis is, however, still on the mental, and the

---

[1] *Mental Development in the Child and the Race,* 2d edition reprinted, 1897, and *Social and Ethical Interpretations,* 3d ed. 1902.

special problem is to determine what sort of a theory of biological evolution is rendered the more probable, when we recognize, together with all the established biological facts and principles, also the principles and facts of the mental life which as psychologists we are bound to accept. In the earlier volumes, we have 'read up,' so to speak, from the individual to his species and to his social group, considered as being also psychological; now we 'read down' from the individual, considered as an organism, to the simpler forms from which he has had his origin — all taken together as constituting an organic whole having a natural history upon the earth.

Looked at in this way, the papers which follow are seen to have the unity of a common purpose, despite the gaps in the presentation of the evolution problem as a whole. They may be treated as each dealing with a narrower question, yet as having reference to the larger problem which may be called *psychophysical evolution* — the evolution of mind and body together. As thus falling into certain groups, the discussions may be classed under the general headings given to the main divisions or Parts of the volume: I., the problem of Genesis as such — some of its main illustrations and data; II., that of the Method of Evolution — involving the determination of the movement, its direction, and the results in which the genetic factors, taken together, actually issue; and finally III., that of Criticism or Interpretation — of finding out the limits, tendencies, termini, and in general the competence of the genetic method in the court of science and philosophy. Genesis, Method, and Interpretation may be taken as the catchwords of such a series of papers whose common motive is covered by the words 'Development' and

'Evolution,' understood in the sense of the distinction made immediately below.

In furtherance of this object the most important distinction, at the very outset, is doubtless that upon which certain great departments of biological science are separated off from one another: that between individual Development and racial Evolution. It is Huxley to whom this distinction of terms is attributed. Development is to be used for the processes of the individual's history from the beginning of its existence in the fertilized egg to its death — the province of fact also set off by biologists by the technical term 'Ontogeny.' The province of racial descent, the tree of connected forms springing from a common stock, together with the entire series of forms which may be represented as branches of the tree of animal life on the earth, this province is that of Evolution, as contrasted with Development — called by the biologists technically 'Phylogeny.' The sciences of Embryology, Experimental Morphology, Physiology, etc., so far as they are genetic, deal with Development; those of Paleontology, Comparative Morphology, etc., deal with Evolution. A still more comprehensive province of research to which the genetic method directly introduces us — whether we deal with the data of mind or with those of life — is that of the interrelation or correlation of these two great spheres, Development and Evolution, with each other. As we shall see later on, certain most vital questions of genetic science come up in connection with such a correlation.[1]

---

[1] No single term has been generally adopted to cover the field of this correlation between development and evolution. The term 'ontophyletic' (determination, concurrence, etc.) might be employed, or the word 'intergenetic,' for cases in which both departments of genetic process are together involved. See the remarks on page 11, note.

Looked at in this way, the problem as a whole — that
of Psychophysical Evolution — requires some preliminary
dissection. Certain distinctions are quite essential, the
more because, if they are too often neglected by biologists
and psychologists alike, it is no doubt partly because they
are dealing respectively with the biological or the psychological, not with both. The first of these distinctions
is that between the two general provinces of research,
Biology and Psychology.

§ 2. *The Psychological and the Biological*

By the psychological I mean the mental of any grade,
viewed from the outside ; that is, viewed as a definite set or
series of phenomena in a consciousness, recognized as
facts and as 'worth while' as any other facts in nature.
The phrase 'natural knowledge' includes knowledge of
psychological facts in just the same sense as that of biological or chemical facts. The occurrence of a psychological change in an animal is a fact in the same sense
that the animal's process of digestion is. And the genetic
explanations which we find it possible to offer, in this case
or that, may draw upon facts of psychology, no less than
upon facts of biology. In the case, for example, of one
animal's recognizing another and being led by this recognition to carry out the act of mating, we have a complex
series of events involving the psychological process of
recognition, joined with that of mating in the production
of one of the great results of nature, and illustrating one
of the principles important to the last degree for the
theory of evolution — the principle of hereditary resemblance. The hereditary traits of the offspring are in this

case what they are because the particular parents mated; but the particular parents mated because one of them recognized the other. The psychological fact of recognition is as necessary to the result as is the process of reproduction. It is a rule, indeed, that *for science all facts are equal.* Such a rule enables us to avoid the recondite question as to which province is to take precedence in this case or that, provided we are dealing explicitly with a problem to which both sorts of fact are relevant.

The recognition of psychological facts becomes especially important in view of the separate way in which analogous questions are often put in the two sciences of psychology and biology respectively. The discussion of the respective spheres of these two sciences turns upon a distinction of points of view. On the one hand, the psychologist as such, and for his science, must aim at the recognition only of the facts which are psychic or mental; that is, of such as are facts to the consciousness *in which they occur.* These alone are psychic, and these belong to individual psychology. So soon as we take up, however, the standpoint of the observer, that of the scientific man who essays to investigate *some one else's consciousness,* or that of an animal, the procedure is now subject to different rules and limitations of observation. To use the terms of a recent distinction of terminology,[1] the facts, while *psychological,* are yet to the observer *not psychic.* The investigation which we set ourselves when we come to discuss psychophysical evolution is 'psychological' in this sense, that is, objective. In the earlier volumes of

[1] Cf. the writer's *Dictionary of Philosophy and Psychology,* art. 'Psychic and Psychological.'

this series we have taken mainly the 'psychic' or subjective point of view, going over to the psychological,
however, when we had occasion to reach interpretations
of a biological or sociological sort. Now the main resource is psychological and biological, the facts of mind
and those of life standing on the same objective footing.
We are now distinctly in the spectator's shoes, observing
facts in the evolution or development of minds and organisms together, at any grade in the series of forms from
lower to higher.

Such a distinction, it is evident, is not possible to the
biologist as such: all of his facts are simply vital as contrasted with the non-vital; there is no question of a subjective over against an objective point of view. So as
regards the relation of biology to psychology we have a twofold distinction: that of the vital as distinguished from the
psychological; and, on the other hand, that of the vital as
distinguished from the psychic. As to the first of these
distinctions, there is a broad truth which may be stated at
this point.

In another place [1] it is pointed out that the entire hierarchy of the sciences is run through by a form of interdepen-
dence as between contiguous departments of research.
The concept of force, when strictly construed, becomes the
touchstone for the differentiation of the sciences. A force
is whatever is present when one stage of a process succeeds necessarily upon another stage. A force always
shows itself in a change, one aspect of process succeeding another; and the passing over of one set of phenomena

---

[1] *Psychological Review*, January, 1902, pp. 57 f., and *Social and Ethical
Interpretations*, 3d ed., 1902, Introduction, § 2; cf. also below Chap. XIX.
§ 8.

into another is necessary to the notion. We are justified, therefore, in finding a force in a set of phenomena only when we are able to find *a continuous process of change taking place in a continuous sort of material.* Given the material of this science or that, — the arbitrarily selected domain of observation, — we may then find forces of which this science may take cognizance when and only when the antecedent is followed by the same subsequent phenomenon, *both in this sort of material.* This, as is said above, holds so long as we restrict ourselves to a limited domain of facts. For example, in an earlier discussion, cited above, the question is that of the definition of the social. We find that various sorts of 'forces,' vital, physical, even chemical, and, by way of climax, changes due merely to the absence of certain usual conditioning limitations — all these have been called 'social forces.' But when we distinguish 'social forces' as 'social producers of change in social material,' we are then able to subordinate all the other loosely recognized agencies, putting them under the heading of 'conditions' — modifying, limiting, and directing conditions — under which the truly social forces operate.

So it is in each science. I have suggested that the term '*nomic*' be applied to such *conditions* considered with reference, in each case, to the true set of forces whose play they condition. The 'socionomic' agencies, forces, etc., using again the same illustration, condition the operation of the forces which are truly social.

Carrying out the same distinction in this present connection, we have analogous results. Psychology finds certain continuous processes of change, certain psychic states antecedent upon certain other subsequent psychic states. These, in accordance with the distinction suggested, we may properly call 'psychic forces,' taken to include whatever we, as psychologists, find it necessary to believe is involved in the conception. The flow of the psychic, we find, however, so soon as we go over to the objective or 'psychological' point of view, is *conditioned* upon physiological processes and functions — those of the brain and other organs. These latter condition — limit, further, direct, inhibit, in any way modify — the flow of the psychic changes. Such conditions are '*psychonomic.*' This term may be used to denote the entire sphere of phenomena which are in connection with the psychological, but which, nevertheless, are not intrinsic to the series of psychic changes as such. Psychology, when considered as the science of mind, in its evolution as well as in its development, — of mind, that is, looked at from the objective point or view, — takes cognizance of the 'psychonomic'; but when considered as a subjective science, as interpreting its own data, it does not ; but, on the contrary, it confines itself to the psychic.

But now, and this is the essential point to remark in our present connection, so soon as we ask the psychophysical question of genesis, — that of the development and evolution of mind and body taken together, — pursuing the biogenetic method, this limitation no longer rises to trouble us. We include all psychophysical facts as such in the definition of our science. Changes in mind and body go on together, and together they constitute the phenomena. Both organic and mental states and functions may be appealed to in our endeavor to trace the psychophysical series of events as such, since both are objective to the spectator, the scientific observer.

The same relation between the intrinsic and the 'nomic'

arises also to confront the biologist. The term 'bionomic'[1] has already gained currency in biology; it is the science of the relations of organisms to their environment, including other organisms. It was, indeed, by way of generalizing this important distinction of the biologist that the general point of view now under discussion was arrived at. If we bring out what is really the meaning of such a distinction in biology, we are led to distinguish the bionomic forces and conditions, those of the environment in all its varied aspects, from the truly biological or vital. Biological forces, properly speaking, are only those which reveal themselves in *vital changes*. The forces of the environment serve to condition, to limit, to direct, the operation of what is truly vital, but they cannot themselves be called vital. They are 'bionomic.'

This distinction on the side of biology is, in the writer's opinion, of considerable importance. Only by recognizing it can general biology develop as an independent science. Vital antecedents of vital changes, — always phenomena of vitality, — these are the matters of biology. Other phenomena may intrude upon the vital, and the morphological changes which become vital may be due in the first instance to such intrusion; but it is only as thus directing vital processes, not as themselves having a claim to be called vital, that these things have significance for the science of life.[2]

While it is true of the enviroment, yet it is not necessary, as has been intimated, to treat the psychological as being bionomic with reference to life, although for the biologist,

---

[1] Suggested, I believe, by Eimer.
[2] Cf. the remarks on Natural Selection, Chap. VIII. § 7. And see the further discussion in the chapter already referred to (Chap. XIX.).

as such, this is an open question, yet it is a gain of no little importance that such a question may be set aside. The equality of facts becomes our rule so soon as we make the problem of evolution a psychophysical one. We may legitimately use such a combination of the mental and the purely vital as that cited above — the case of recognition-marks — without stopping to inquire in what sense a mental fact, such as recognition, can have causal value in the determination of purely physical characters in the next generation. That may be discussed in psychology, or in biology, and it must be discussed in genetic philosophy; but in a department in which the psychophysical as such is the type of phenomenon expressly taken up for examination, the divorce of the two, and even the recognition of a dualism between them, is unwarranted.

§ 3. *Psychophysical Parallelism*

With the general understanding now arrived at, we may take a preliminary survey of the field in the light of certain current hypotheses. Among these is what is known as 'psychophysical parallelism.'

This principle, as ordinarily stated, supposes a thoroughgoing concomitance between the two terms of the psychophysical relation, mind and body. It states the general fact that certain changes in the organic, in those brain and nerve processes with which consciousness is associated, are always accompanied by changes in consciousness, and also, that this last is a statement which can be converted — so that it is also true that all changes in consciousness are accompanied by organic changes, in the brain and nerves.[1]

---

[1] Much embarrassment is likely to arise from confusion of terms, as the problems of psychology and those of biology are brought into closest union.

This principle, now made the assumption of experimental work in psychophysics, would seem to involve, and also to be supported by, certain other formulas which are a part of general scientific procedure.

First, the principle of *equal continuity*, to the effect that there can be no breaks in either series of changes, the brain changes or the conscious changes, without a corresponding break in the other; in other words, if one of the series be continuous, the other must be continuous also. This is referred to again below.

Second, the principle of *uniformity*, to the effect that the sort of modifications which are associated one with another in brain and mind are always the same; that is, if a certain brain process be correctly hit upon as essentially associated with a certain conscious state, then the concomitance of these two terms may be looked for

The word 'parallelism' has been in use for a long time in philosophy and psychology for the relation of body and mind, and it is impossible to discard it in the present connection. Yet the biologists have used it also for the relation expressed much better by the term 'recapitulation' (so Cope), and also for the 'parallel' or concurrent direction of development and evolution as determined by any given influences. In order to avoid such confusion, I shall use 'parallelism' for the psychophysical relation as explained immediately above, 'concurrence' for the determination of development and evolution in a common direction, and 'recapitulation' for the relation of the two series whereby development reproduces evolution. The term 'concurrence' in such a sense is suggested in my *Dictionary of Philosophy and Psychology*, art. 'Organic Selection.' For further illustration, we may say that psychophysical parallelism is both 'individual' and 'racial,' and is found to illustrate both recapitulation (from the fact that the mental and organic series in development recapitulate respectively those of evolution, that is, if either series does, the other, by the law of parallelism, must also), and also 'concurrence' (as a fact, *i.e.*, based on researches which show that evolution follows a course first marked out by individual development). Recurring to an earlier suggestion (p. 3, above) we may note that all three of these conceptions are 'intergenetic,' or 'ontophyletic' (the former term being the one which I prefer, and shall use).

on all other occasions of the occurrence of either of them.

This formulation is, it is easy to see, absolutely necessary to any science of psychophysics at all. The theory of the localization of functions in the brain, for example, assumes it. For if vision has its seat in the occipital region at one time, it may be assumed that at another time a lesion of that centre will interfere with vision, or that certain troubles of vision may be taken as evidence of lesion in that centre. If these expectations be not fulfilled, the only alternative left open to the investigator is to believe that the first determination of the seat of vision was erroneous.

This requirement alone — the demand for uniformity in the facts with which psychophysics has to deal — is itself sufficient to justify the acceptance of psychophysical parallelism as against all the theoretical objections that may be and have been urged. As a rule of scientific procedure, it is a necessary assumption. To deny it is to say antecedently to the actual examination of the facts which it claims to formulate, that a natural science of the individual as a whole is impossible; for any formulation of the facts must proceed upon the assumption that they have sufficient regularity and uniformity to allow of formulation.

Third, the principle must be a universal one, if it be valid at all; which means, that wherever we find a series of phenomena which are psychophysical, this principle of parallelism has its application. This leads to the necessity of extending the application of parallelism from the sphere of development to that of evolution, as those terms are distinguished on an earlier page. This may now be explained in a little more detail, after which certain applications of the two preceding principles — continuity and uniformity — may also be indicated.

§ 4. *Psychophysical Parallelism in Evolution*

I have said that the principle of parallelism is universal, that is, that it is applicable to all instances in which the facts on either side are of the order indicated by the term 'psychophysical.' The great spheres in which such a truth would have bearings are the two covered by individual life history, from life to death — ontogeny, the sphere of development — and the race history of a species, or of all life upon the earth considered as showing a series of forms connected by links of progressive descent — phylogeny, or evolution. These two forms of parallelism we may call respectively 'individual' and 'racial.' Earlier discussions of parallelism have had reference largely to the former sphere, and the question has come up for the most part in connection with theoretical discussions of the relation of mind and body. Furthermore, the concomitance of development of mind and body in the individual has had recognition, and its illustrative value for the topic of evolution has been written about — though not at all adequately.

The corresponding racial application of parallelism in the sphere of evolution has not, to the present writer's knowledge, been explicitly made ; that is, as going necessarily with an individual parallellism, as part of an intergenetic conception. And yet the two cases must necessarily go together, and no final formulation is possible of the relation of conscious changes to bodily changes which does not have direct application in both these spheres, and indeed the same application in both

of them. For in psychology, as in biology, the race
series is but a continuous line of individual generations,
and to ask the question of the race is but to ask whether
parallelism holds for any given number of generations
of individuals wherever chosen in the line of descent —
this provided we admit that descent is by some form of
continuous hereditary transmission.

The questions which arise about heredity, however, do not
trouble us, seeing that they are not within the domain of
strictly psychophysical inquiry, except in so far as our
theory must explain the inheritance of both physical and
mental characters to the same degree. For example, the
question of the 'continuity of germ plasm' may be de-
cided one way or the other — either for or against the
actual continuous transmission of an identical substance —
without raising the question of a corresponding transmission of anything psychological. For if there be breaks in
the psychological series at those nodal points at which generation succeeds generation, there are also, by the principle of
equal continuity, discussed above, breaks also in the psychophysical, and we may find the psychological series beginning
again at the appropriate point in the development of the
organism of the new generation—the point at which the
psychophysical again begins. In other words, the advantage
gained from the psychophysical point of view is that if there
be apparent gaps in one of the series, we may either assume them filled up by theoretical parallelism with the other
series at these points, at which it has no gaps, or we may
— if we deny continuity to either — make gaps in the
second series in correspondence with the gaps found in
the first. We have in any given case, in short, either a
psychophysical fact, or we have not: if we have, then

either series is sufficient to carry us over the critical point; if we have not, then the break in one series is sufficient evidence of a corresponding break in the other also. The principle of parallelism assumed, we claim once for all the right *to neglect the relation of the two terms, mental and physical, in all circumstances whatsoever*.[1]

On this way of conceiving the scientific inquiry, we may proceed unhampered by the problems which trouble the philosopher. We do not have, for example, to adopt Professor Lloyd Morgan's theory of 'metakinesis,' postulating something quasi-mental to fill out the breaks in the psychological series, at points at which we have no sign of the presence of consciousness. Nor do we have to embrace the idealistic theory of knowledge of his critic, Professor Karl Pearson, to do away with a troublesome brain substance, which at times becomes uncomfortably prominent. The formulations of 'shorthand,' to use Pearson's phrase, may be made for both series together. This is required for such a problem as that of evolution. We do not have one series of genetic forms, the mental, evolving under shorthand formulas of its own; and another series, the organic, doing the same thing under different formulas. On the contrary, the two sets of facts really go together in the one set of formulas. This is what I am arguing for. We often find it necessary to use the mental facts as antecedents of the physical facts, often the physical as antecedent of the mental, and again, often the psychophysical as antecedent of either or both. This possibility is presented with more detail on a later page (Chap. IX. § 3).

The twofold application of parallelism, considered as an

---

[1] Cf. the further remarks on the construction of heredity below, Chap. XVI.

assumption of psychophysical research, may be represented by the accompanying diagram. The two vertical lines (M, B) represent the two series in the evolution of the race-forms of organisms — the dotted line (M, mind) being the mental, and the solid line (B, body) the physical. Across these at any point we may draw similar horizontal

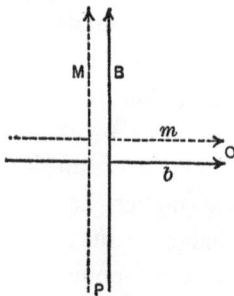

lines (m, b) representing individual development in any given generation; these are also, of course, dotted (m) and solid (b). The full theory of parallelism requires, not only that we make the two horizontal lines parallel, — the ordinary application in ontogeny (O); but that having gone so far, we must also draw the two parallel vertical lines — the application in phylogeny (P). At whatever point in the line of descent we apply the principle to individual development, we must perforce raise the corresponding genetic questions about the evolution which has led up to the birth of such individuals at that point. And the series of 'shorthand' formulas, laws — in the prosaic equivalent of everyday science, 'results' — at which we arrive, must involve the three great problems represented by the four lines: parallel development (the relation of m to b), parallel evolution (the relation of M to B), and intergenetic correlation (the relation of mb to MB). Furthermore, when we recognize in places the absence of the facts we should expect, — apparent breaks in either one of the lines, — we may resort to the resource of using the corresponding facts from the parallel line at the same level, and even those from the analogous line of the other pair of parallels, so far as there are known facts in the particular case which lend themselves to such procedure.

The philosophical problem is illustrated by a similar diagram in the section below already referred to (Chap. IX. § 3). In that connection it is a question of reaching an interpretation in a statement in which the independence of the two series is in so far denied — the representation being that of a possible single line which can be substituted for the two in their development. A similar philosophical problem is open also in the matter of evolution. Such an interpretation in either province, it is maintained in the later place, is not possible to the scientific inquirer as such, although it may be possible to arrive at a philosophical explanation of the dualism of mind and body. In our present connection the urgent need is in another direction — to hold a level balance and give each side its due. It is an equally embarassing thing for the scientific inquirer in one case to be a monist to such a degree as to deny one of these lines altogether — the mental in favour of the physical 'shorthand,' or the physical in favour of the mental — and in another case to insist upon the separateness of what nature shows us always joined together, to the extent of refusing to use facts from one of the series to illumine and even to explain facts in the other.

This last-mentioned attitude is especially to be condemned in the discussion of genetic questions —those of development and evolution. Here the question turns upon the genetic antecedents of a given fact, be it function, act of behaviour, mental state, instinct, or other, in this organism or that. No tracing of genesis is possible, of course, except by actual observation of the facts under which the phenomenon in question arises. Now to refuse

c

to see certain of the antecedents because they are not physical, or because their physical counterpart is not known, or the reverse, that is, to observe only the mental antecedents of the fact in question, is suicidal to genetic science. It not only omits facts from its formulations, but it actually does violence to facts. For with the omission goes the commission of the positive error of arriving at an interpretation which is false.[1]

A remarkable instance illustrating the necessity of recognizing both orders of facts is to be found in the theory — and the history of the theory — of 'warning colours.' As preliminary to the theory there is the fact of coloration, which is distinctly physical. The question is as to its origin. The theory holds it to be due to the warning given to other individuals that a particular colouring is distasteful or poisonous. Now, in order that this warning be given, the biologists tell us there is necessary a certain education of the hostile individuals. The creatures have to learn the meaning of the coloration; and this learning involves profiting by experience. If each creature made the same experiment each time instead of profiting by his own experience, or if each had to learn for himself, instead of profiting by the experience of others through imitation, etc., of course there would be no utility in the colouring as giving warning. Here is as distinctly a mental process involved as any one might cite. To refuse to recognize it would be to throw away what is generally recognized as the true theory of these cases of coloration.

---

[1] Cf. the criticism of Professor James' theory of the separateness of the psychological and physiological 'cycles' in my *Social and Ethical Interpretations*, Sect. 42. See also the further discussion in Chap. XIX. on 'Genetic Modes,' below.

The action of natural selection, I may add for completeness, secures the survival of the insects so coloured, seeing
that being warned, their enemies let them alone. The
possibility of the evolution of the definite coloration turns,
in fact, upon this series of psychological processes.[1]

[1] On this particular topic see Professor Poulton's *Colours of Animals.* A
recent concise statement of the facts by the same writer is to be found in the
Dict. of Philos. and Psychol., arts. 'Warning Colours,' and 'Mimicry.'

CHAPTER II

COMPARATIVE CONCEPTIONS

§ I. *Recapitulation*

THIS way of looking at the two spheres of development and evolution, as involving an application of the one principle of parallelism, carries with it certain consequences of considerable interest. In the first place, it requires us to carry over into the genetic treatment of psychology the same thorough-going genetic point of view which evolution postulates in biology. And with this goes the question of the application to the facts of the one of the principles already established for the other. The great law of recapitulation at once comes to mind, the law with which we have been having considerable to do in the earlier volumes of this series. If we hold that mind and brain processes are parallel as well in the species as in the individual, and also hold that the brain series in the individual's development recapitulates in the main the series gone through by his species in race descent or evolution; then it follows that the law of recapitulation must hold also for the mental. This has been recognized and the limitations of it have been pointed out in Chap. I. of *Mental Development*:[1] and a further general application of the idea to

[1] In that place, under the topic 'Analogies of Development,' the main 'epochs' of growth in each series, which illustrate the recapitulation of evolution by development, are briefly indicated. The following quotation from that work presents a preliminary statement of the general thought which is now worked out explicitly and with greater detail in these pages : " Assuming then that there is a phylogenetic problem, — that is, assuming that mind has

the social life is worked out in the *Social and Ethical Interpretations*. The further extensions which the theory of biological recapitulation has recently undergone, prominent among which is the theory that regression shows itself first in individuals and afterward in the same directions in the species, should also find confirmation if they be true, or the reverse if they be not true, from the facts of genetic psychology.[1]

§ 2. *Natural and Functional Selection: Plasticity and Intelligence*

Apart from the question of recapitulation, which I have given this prominence both from its general character and also from the fact that it has survived very considerable criticism in recent literature, we find certain points

had a natural history in the animal series, — we are at liberty to use what we know of the correspondence between nervous process and conscious process, in man and the higher animals, to arrive at hypotheses for its solution; to expect general analogies to hold between nervous development and mental development, one of which is the deduction of race history epochs from individual history epochs through the repetition of phylogenesis in ontogenesis, called in biology 'Recapitulation'; to view the plan of development of the two series of facts taken together as a common one in race history, as we are convinced it is in individual history by an overwhelming weight of evidence; to accept the criteria established by biological research, on one side of this correspondence, — the organic, — while we expect biology to accept the criteria established, on the other side, by psychology; and, finally, to admit with equal freedom the possibility of an absolute beginning of either series at points, if such be found, at which the best-conceived criteria on either side fail of application. For example, if biology has the right to make it a legitimate problem whether the organic exhibits a kind of function over and above that supplied by the chemical affinities which are the necessary presuppositions of life, then the psychologist has the equal right, after the same candid rehearsal of the facts in support of his criteria, to submit for examination the claim, let us say, that 'judgments of worth' represent a kind of deliverance which vital functions as such do not give rise to." (First ed., pp. 14 ff.)

[1] See also Chap. XIII. § 5, on 'Concurrence and Recapitulation.'

at which what may be called 'comparative' questions
arise. By this I mean points at which the interpretation of a series of facts has been fairly made out, or the
facts at least formulated, on one side of the two parallel
series, and we may properly ask how far the same or
an analogous interpretation or formulation is possible on
the other side. For example, the law of natural selection
from spontaneous variations, which makes use of the
criterion of fitness or utility only — what can we say of
mental evolution from this point of view? We have
here to apply a biological conception directly to the
mental. Again, in development the mind seems to progress by a certain function of selection by which it brings
itself into better accommodation to a complex mental and
physical environment. Here is a certain formulation on
the mental side, a function so well recognized that the
criterion of consciousness in an organism is often said to
be the exhibition of a 'selective' reaction. What now
can the biologist do with this in his theory of organic
development?

Both of these instances are enlightening for the derivation of what I am calling 'comparative conceptions.'
The union of the two seems to require that the brain
variations by which evolution proceeds be of such a sort
that their very utility — that for which they are selected — is in the line which mental development by a
selective function in each generation acquires. Now, in
fact, we find this requirement fulfilled in the evolution,
by variation and natural selection, of increasing *plasticity*
of nervous structure. By this, not only does it appear
that mental evolution progresses by variation, keeping
pace with organic (a correlation in evolution), but also

that organic variation is in a direction which furthers accommodation, or 'educability,' by mental selection in individuals (a correlation in development). In the latter respect, the utility of plasticity, as permitting mental development and selective accommodation, is so great that it outweighs in evolution all other biological characters, and as embodied in the brain, becomes the great outstanding fact throughout the ascending series of mammalian forms. Once given the presence of consciousness, with its methods of psychological accommodation, and it carries with it organic adjustment; while the operation of variation with selective survival tends to perfect it.

Moreover, in psychophysical accommodation we have the problem of selection set in a new form inside of the functions of the organism itself. Consciousness, mind in any form, is a character; its functions are always psychophysical in their operation. How then can consciousness select without violating parallelism ? In answer to this it may be pointed out that accommodation is another comparative conception ; for it is only by an application of the operation of natural selection in the form of survival from overproduced functions, such as movements, dispositions, etc., that consciousness can effect selective adjustments ; that is, there is no other way short of a miracle. This in general outline is the conception of functional selection.[1] So a completed view of psychophysical accommodation requires (first) natural selection, operating upon (second) variations in the direction of plasticity, which allows (third) selective adjustment through

[1] Worked out in *Mental Development* on the basis of the theory of 'surplus discharges' of Spencer and Bain. See also the recent work on *Animal Behaviour*, pp. 163 f., by Professor Lloyd Morgan.

the further operation of natural selection upon the organism's functions.

So we have certain comparative conceptions : variation with natural selection, consciousness or intelligence with plasticity, and accommodation by functional selection. They are comparative in the sense which is now occupying us, that is, psychophysically ; since the meaning of any one of them is not exhausted in its application to either mind or body, without appeal also to the other of these two psychophysical terms.

For example, the meaning of the evolution of the brain cortex, with the extreme plasticity which is its main character, through the entire line of mammalian descent, can be understood only when we recognize the evolution of intelligent endowment which accompanies it; and the method of the selective function of consciousness can only be understood, in my opinion, in the light of the method of survival by selection from overproduced variations, which is the method of natural selection.

§ 3. *Correlation of Characters*

Another highly interesting comparative conception is that connoted by the biological term 'correlation.' [1] This idea covers the fact that certain characters of the organism are correlated with, connected with, or in other regular ways related to, certain other characters, in such a way that modifications or variations of the one are accompanied by changes in the other also. This is true, not only of those characters in which it is difficult to determine the precise function, and of which, therefore, the definition itself is difficult and uncertain, since it involves

[1] See instances given in Chap. XIV. §§ 1-3.

others as well; but also for those which are remote from one another, as, for example, the internal glands whose secretions are found to be in some obscure way correlated with general conditions of the organism. This principle, when once formulated, will undoubtedly be important; but as yet no exact laws of correlation have been made out. Yet it is quite allowable to say, in the same general sense, that psychological and psychophysical correlations hold. The psychophysical relation is itself a correlation having many special illustrations, such as the correlation between plasticity and intelligence, that between the fixity of nervous processes and the corresponding impulses, instincts, etc. Indeed, the psychophysical question, when put in a particular case, is really one of determining the correlation of the characters which consciousness shows with those of the organism as such. And it follows that, wherever a mental character enters into the complete carrying out of a physical function, it must have its place assigned to it, according to what has already been said, in the complete determination of the genetic significance of that function.

Furthermore, we may say that no physical character which has mental correlations is completely understood until these latter are exhaustively determined, and also that no mental character escapes physical correlation. Recent research in the psychological and physiological laboratories is establishing many such psychophysical correlations: that of emotion with motor processes, of attention, rhythm, and the time-sense with vasomotor changes, that of mental work with nervous fatigue, and so forth through all the main problems of this department. All this affords, in so far, at once illustration and proof of the general formula of psychophysical parallelism.

§ 4. *Psychophysical Variations*

Allusion has been made to the great biological topic of variation as embodying a conception common to the two series — mind and body. It is only recently that the theory of selection from congenital variations has been brought over into psychology. Formerly the idea of hereditary transmission of the results of mental education was simply assumed. But the failure of that idea in biology has led to the revision of the facts with an equally pronounced verdict against it in psychology also. Beginning with certain brilliant independent examinations of the question, notably that of W. James,[1] the theory of mental variations has come in to account for the evolution of mind in strict correlation with that of the organism. We find not only the correlation of intelligence with plasticity, as pointed out above, but also many other correlated details which the psychophysical processes actually exhibit. This means that natural selection has worked upon correlated *psychophysical* variations — not upon organic variations merely. In other words, *it has been the psychophysical, not the physical alone, nor the mental alone, which has been the unit of selection in the main trend of evolution*, and Nature has done what we are now urging the science of evolution to do — she has carried forward the two series together, thus producing a single genetic movement. It would have been impossible for mind to develop by selection with reference to utilities for which the necessary organic variations were not present; and so also it would have been impossible for the organism to evolve in ways which the consciousness of the same animal forms did not

---

[1] *Principles of Psychology*, II. Chap. XXVIII.

support and further. There could not be independence; there must be correlation.

This is illustrated by several of the facts and principles pointed out in the following pages. It has been argued in the earlier volume on *Social and Ethical Interpretations* (Chap. VI.) that emotion shows a development from an 'organic' to a 'reflective' or intelligent type, which latter, however, utilizes in its expression the same organic processes as the former; and it is there stated that this could have come about only by the sort of correlation now under discussion. Only those reactions of the organism, selected for their utility in offence, defence, etc., would survive, which could either be actually used for the higher purposes of mind, or which, at least, did not stand in the way of the exercise of the higher functions. Both of these possibilities are realized, and in some cases we find the presence of vestigial 'expressions,' now harmless although no longer useful; while in other cases the original reaction has been modified to serve the new purpose. In certain cases, also, these vestigial reactions or dispositions are, in some degree, disturbing factors to the possessor of the new functions.[1] It is argued below (Chap. VI. § 1) that both the instinctive or reflex and also the intelligent performance of a given function may coexist side by side, each having utility and each preserved for its utility — an additional resource thus being given the possessor in coping with complex circumstances. In this case, there has been a selection of variations toward the plasticity which the evolution of intelligence demanded, together with the growth of the apparatus of voluntary movement, while at the same time the fixed connections requisite to the reflex or instinctive per-

[1] So blushing, as is maintained in the work mentioned, Sects. 134 f.

formance of the same functions have not been disturbed, the same apparatus being so modified, however, as to serve the two utilities in question more or less independently.

Another case of interest from the psychological point of view is that of the genetic interpretation of the function of imitation, itself quasi-instinctive, or impulsive, in relation to other mental and organic functions. As I have argued in detail in *Mental Development*, considered genetically as a type of reaction, imitation involves reference to an end or 'copy,' which is the prime characteristic, also, of intelligent action; but it is held down to a definite psychophysical process, called the 'circular' process, whereby the copy is reinstated by the act of imitation. For example, my parrot has just learned to say 'Hulloa' imitatively. He learns to pronounce this word just as an intelligent child would learn to do it; but he cannot vary, modify, or inhibit it, nor exercise selection in the manner of his doing it. His act seems to lie, therefore, as type of function, midway between the congenital instinct and intelligent selective action. The present writer considered this function to be probably a case in which natural selection has put a premium upon the acquisition of adjustments which would keep a creature alive and give the species time to acquire the congenital mechanism for performing the same functions — illustrating what is called, below, 'organic selection.' Imitation would, thus considered, in many cases aid the development of instincts; in all cases, that is, in which the instinctive performance would, by reason of promptness, accuracy, etc., be of greater or of additional utility. But about the same time Professor Groos published his theory of play in a work in which imitation is held to have just the opposite genetic relation to instinct and intelligence.[1] According to Groos, the imitative performance, by reason of its character as presenting a certain degree of selective learning and accommodation, tends to supplant the fixed reactions of instinct, and so to put a premium on variations toward the plasticity required by increasing intelligence. It now transpires that Professor Groos and I are able to accept each the other's position, and so reach the common view that it depends upon the exigencies of the particular adaptation required by the animal species as to what a particular imitative reaction means. If an imperfect instinct is in the way of development for a marked utility, imitation, by supplementing it, would undoubtedly aid its survival and evolution in the way indicated above. Yet, on the other hand, if an instinct is in process of decay, — or if the conditions make its decay desirable, — Professor Groos' principle would then come into operation. The imitative performance would represent a form of variation which would be in the direction of the plasticity of intelligence, and creatures would be selected who performed the function imitatively, until further variations toward plasticity were forthcoming. In either case, and especially in both cases working in nature together, we have a clear illustration of the sort of psychophysical 'togetherness,' so to speak, the indissoluble correlation, into which the organic and the mental are welded in the process of evolution.

The fact of correlated variation, moreover, is to be carried over to the relation between organic and mental variations *in different individuals*. Many instances are known which prove it; that they are not more numerous is due,

[1] *The Play of Animals*, Eng. trans. Cf. the notice of that work below, Appendix C.

I think, to the neglect of the recognition of it in seeking genetic explanations. For example, sexual selection requires correlation between the organic characters of coloration, etc., in the male, and the mental apprehension and the sexual impulse of the female. So the evolution of infancy requires correlation between the physical helplessness of the young, and the maternal instinct and affection of the mother. In the evolution of gregarious life we find a vast system of correlations of physical characters, — expressions, attitudes, behaviour in general — which are interpreted and responded to psychologically by other members of the group; these physical and psychological characters together make up the psychophysical equipment of the individuals for their common life. In a later place (Appendix C) the possibility of correlation between mental characters and sexual variations is pointed out in connection with Pearson's theory of 'reproductive selection.' It is remarked also in the same place, that one great form of isolation, that due to social barriers which create segregation and preferential mating, and so effect physical evolution, is not noticed by Romanes in his description of the different forms of isolation; here there is involved a correlation between the mental functions embodied in personal choice, social convention, law, etc., and colour of skin or other physical characters which either attract or repel. The theory of 'secondary sexual characters' in man and woman extends to mental traits, and points out correlations not only between many characters in the same individual, but also between these of individuals of the opposite sex.

The theoretical importance of this sort of correlation appears more fully when we look closely at what it involves. In the first place, negatively: if it be true that the unit of selection in evolution is often a psychophysical function or character, it may be only the failure of psychophysical observation which prevents the genetic explanation of a series of organic changes. The search for utilities should be extended to the mental sphere. The larger utility of the psychological or the psychophysical may be the key-note in a case of survival; and the failure to discern this utility may block our scientific progress. So it is, for example, in appreciating many forms of play. If we adopt the 'practice' theory, which holds that play is a means of preparation, through preliminary practice, for the strenuous specific activities of adult life, we must recognize that it is often mental practice — in accommodation, judgment, social adaptability, etc., — or the training of mental functions, which is the critical utility; and that to understand this utility is at once to secure an application of natural selection, where otherwise, from the purely organic point of view, no adequate ground of selection would have been discoverable.

While so much is true negatively, the matter has also a positive aspect. The actual construction of a view regarding a particular function or character can often be arrived at only by weighing the psychological facts. So in the case of the function just cited, animal play. There were earlier theories of play. The 'surplus energy' theory of Spencer was generally held, despite the quite valid criticism that it had the negative defect pointed out immediately above : no adequate selective utility attaching to play was involved so long as it was thought to be due to discharges of surplus animal vigour only. The consequence was that play — together with the whole province of art, which is thought by many to have its roots in the play-impulse — was looked upon as a by-product, an unjustified remainder, not due to selection at all, and subserving no utility in the economy of the genetic processes of evolution. Now, thanks to the illuminating works of Groos,[1] developing the scattered hints of others, we discover the psychological and sociological utilities of play, which supplement its biological utility in the practice theory; and the whole is an important contribution, not only to the body of evidence for Darwinism, but also to the psychophysical interpretation of evolution. Play and art are now no longer luxuries for the rich; they are necessities as well for the poor — to speak in terms not entirely figurative.

Indeed, in this conception of correlated variation many of the mysteries of evolution are pooled. The position taken above, and elaborated in the later chapter, to the effect that the conditions which are 'nomic' to a genetic movement are to be carefully distinguished from the forces intrinsic to the movement, avails to indicate the capital importance of the fact of variation in mind and body together. Natural selection is in itself a negative principle, a 'nomic' or directive condition; heredity is a principle of conservation in so far as it is specifically and only heredity; and the remaining foundation stone of the entire evolution structure, variation, remains the point of direct and emphatic importance. In it the intrinsic vital processes must exhibit themselves. It is by variation that the materials of selection arise, it is the character of variation that must decide the question of determination — the issue between vitalism and the opposed views. Here, in the opinion of the writer, much of the great biological work of the future is to be done. Witness indeed, the

---

[1] *The Play of Animals*, and *The Play of Man*.

researches already carried out by statistical methods, aiming to determine the actual facts as to whether variations have an intrinsic drift in certain directions, or whether the appearance of such a drift is entirely due to processes of selection within or without the organism. The newer view, which holds that species originate in abrupt or 'sport' variation, called 'mutation,' strikes at the very foundations of the Darwinian conception — that is, if mutation be considered not merely an exceptional case but the normal mode of the origin of species.[1] We may accordingly go a little more fully into the requirements of a theory of determination.

[1] The appearance of the new journal, *Biometrica*, is witness to the vitality of the movement to treat biological phenomena, notably variations, by exact statistical and mathematical methods. Cf. the summary articles by Davenport and Weldon on 'Variation,' and those on 'Natural Selection' and 'Mutation' by Poulton, in the *Dict. of Philos. and Psychol.* On mutation see De Vries, *Die Mutationstheorie* (1901). A summary article by De Vries is to be seen in *Science*, May 9, 1902. It seems to the present writer a very long step from the observation of single cases, admittedly very rare, of the persistance of abrupt variations, to the theory of the 'Origin of Species by Mutation.' For an able negative criticism of De Vries' work, written from the point of view of recent statistical 'biometric' researches, see Weldon, in *Biometrica*, Vol. I. Part 3, pp. 365 ff.

D

CHAPTER III

THE DIRECTION OF EVOLUTION

§ 1. *Genetic Determination: Congenital and Acquired Characters*

THE problem of determination, in its varied aspects, is no more than the problem of the method of evolution; hence the attention given in the following pages to this topic in both its phases, that of evolution and also that of development.[1] As an intergenetic conception it takes form as follows: first, what determines the development of the individual, both bodily and mentally, or in a word, psychophysically? — second, what determines the evolution of the species, in both the same two phases, that is, psychophysically? — and third, how can these two forms of determination work together so that *race determination is 'concurrent' with individual determination?* It is only when all three phases of the problem are held together that the extraordinary complexity of the data comes fairly out. The data of fact and of principle resting upon formulations of fact, as they appear in the present state of knowledge, may be presented somewhat as follows, while the later chapters may be looked to for treatment of various of the subordinate topics which fall under the larger heading.

First, individual development seems to take place by gradual accommodation to environment on the basis of

[1] See especially Chaps. X. and XVII. below, on 'Determinate Evolution' and 'Selective Thinking.'

the congenital hereditary impulses which characterize the species. This is true both of mind and body; and the relation of the respective functions of mind and body varies with the place of the creature in question in the scale of life—with what we may call, technically, its 'grade.' The correlation already pointed out between increasing plasticity of the nervous system and increasing mental endowment holds as we ascend from a lower to a higher stage. We accordingly have an increasing dependence upon accommodation of the mental type as we ascend higher in the scale. The range of *possible accommodation of the organism of a whole becomes, therefore, wider and its congenital impulses less fixed* as evolution advances; there is constantly less dependence upon definite heredity, and more upon the inheritance of a general mechanism of accommodation of a psychophysical sort, as we ascend the animal series.[1] Recognizing progress in progressive accommodation, with plasticity of mind and body, as the direction in which evolution is determined, we may set that down as the first point in our argument. The method of accommodation, its progress by the selection of adaptive movements and thoughts from overproduced cases by trial and error, may be left over for the present.

Second, it follows that the distinction so long dominant in biology between 'congenital' and 'acquired' characters, cannot be sharply drawn. All characters are partly congenital and partly acquired. The hereditary impulse is at the start in each case a rudiment (*Anlage*), which is to develop into what the environment, within which its native tendencies must show themselves, may permit it to become.

[1] Professor Ray Lankester has paralleled this with the advance in size and complexity of the mammalian fossil brain (*Nature*, LXI. p. 624).

This impulse is definite enough in many cases, where the conditions do not require accommodation and modification; but where these demands are urgent upon it, it is surprising what transformations it may undergo. Recent results of embryological and morphological research have proved this so clearly that a school of biologists, called by Delage ' Organicists,' [1] has arisen, who place the emphasis in all evolutionary change upon the necessity of the organism, and of its particular organs, to become what they are stimulated to become under the stress of the environment, or, failing to meet these requirements, to die in the attempt. This suggests an important modification of the strictly ' Preformist' view, made extreme in the earlier writings of Weismann, according to which the accommodations of the individual organism are of no importance, being simply the unfolding of what is preformed in the germ. For even if we admit, as we may, the non-inheritance of acquired characters, we may still hold the general view of the organicists, and also maintain that the hereditary impulse becomes more and more *unformed*, rather than preformed, as we advance in the animal scale; each succeeding generation through its own development, in its own life history, making more of the essential accommodations which give it its generic and specific characters. This Weismann has lately in part recognized, in his theory of ' intra-selection ' built up upon the views of Roux.

Third, if these be the safe results of research in the sphere of development, we then have certain additional

---

[1] ' The Organicists oppose [to other theories] the combination of a moderate predetermination with the continually acting and necessary forces of the environment, which are not simple conditions alone, but essential elements in the final determination.' (Delage, *La Structure du Protoplasma*, p. 720. Delage's personal views are cited in Chap. XIII. § 3, below.)

guiding indications for the problem of determination in the sphere of evolution. The evolution series becomes in its hereditary character more and more indeterminate, more and more indefinite, in respect to what will be produced by the union of the heredity impulse with the conditions of development of the successive generations of individuals. There is a general direction of progress, secured by the natural selection of variations in the direction of the plasticity which increasing accommodation requires; but the utility of this shows itself in the decay of special congenital functions and the increased freedom of the organism in working out a career for itself. Thus there is secured a *blanket utility*, as it were, a general character, through the operation of natural selection, which progressively supersedes and annuls many special utilities with their corresponding adaptations, while, at the same time, other special functions having special utilities are given time to reach maturity by variation and selection.

§ 2. *Genetic Determination: the Factors*

The truth of this position regarding the direction of evolution appears from the detailed explanations by which the two leading positions of this work are supported in the following pages.

Of these two positions the first is that of Organic Selection, explained and applied with considerable repetition below. This position is the general one that it is the individual accommodations which set the direction of evolution, that is, which determine it; for if we grant that all mature characters are the result of hereditary impulse plus accommodation, then only those forms can live in which *congenital variation is in some way either 'coincident' with, or correlated with* [1] *the individual accommodations which serve to bring the creatures to maturity.* Variations which aid the creatures in their struggle for existence will, where definite congenital endowment is of utility, be taken up by the accommodation processes, and thus accumulated to the perfection of certain characters and functions. The evolution of plasticity, on the other hand, could only itself have taken place by the coöperation of accommodation using the variations toward plasticity already present at each stage, and thus saving and developing such variations. This gave an ever higher platform of variation from which steady refinement of plasticity and its accompanying intelligence was all along possible. *Organic selection becomes, accordingly, a universal principle, provided, and in so far as, accommodation is universal.*

Accommodation, therefore, when all is said, is a positive thing, a vital and mental functional process supplementary to the hereditary impulse. It must be considered a positive factor in evolution, a real force emphasizing that which renders an organism fit; whereas natural selection, while a necessary condition, is yet a negative factor, a statement that the most fit are those which survive. If it be true that those variations which can accommodate, either very much or very little, to critical conditions of life are the ones to survive, and that such variations will be accumulated and will in turn progressively support

---

[1] See below, Chap. XIV., for treatment of the distinction indicated by these phrases. ' Coincident variation ' was suggested by Professor Lloyd Morgan : cf. below, Chap. XI. § 1, and Lloyd Morgan's *Animal Behaviour*, p. 37.

better accommodations, then it is the accommodations which set the pace, lay out the direction, and prophesy the actual course of evolution. This meets the view of the Lamarckians that evolution does somehow reflect individual progress; but it meets it without adopting the principle of Lamarckian inheritance.

The second general position advocated, on the basis of facts, in the following pages is that of Social Heredity,[1] or Social Transmission, with Tradition. This too falls into place in our general theory of determination. If accommodation is a fact of real and vital importance, then some natural way of regulating, abbreviating, and facilitating it would be of the utmost utility. If animals were left to constant experimentation each for himself, they would die, as we have said above, before they made much development. We find that an important function of consciousness is that it enables them to profit by experience. By memory, association of ideas, pleasure and pain motivation, they abbreviate, select, and handle experience to the most profit.

But there also arises an additional resource — and certainly a very important one — by which *they are enabled to profit as well by the experience of others.* So soon as animals can use their native impulses in an imitative way, they begin to learn directly, by what may be called 'cross-cuts' to a desirable goal, the traditional habits of their species. The chick which imitates the hen in drinking does not have to wait for a happy accident,

[1] In the earlier volumes of this series, where the psychological process of acquisition is much in discussion, the phrase 'Social Heredity' is largely used. In the following pages, wherever possible, the expression 'Social Transmission' is employed.

nor to make a series of experiments, to find out that water is to be drunk. The bird deprived of the presence of others of its kind does not learn to perfection its proper song. All the remarkable accommodations of an imitative sort, so conspicuous in the higher animals, enable them to acquire the habits and behaviour of their kind without running the risks of trial and error. Calling this store of habits of whatever kind 'tradition,' and calling the individual's absorption of them and his consequent education in tradition his 'social heredity,' we have a more or less independent determining factor in evolution. For these accommodations are the cream of the needs of life, they represent the essentials of education, the *sine qua non* in an animal's equipment; so the accommodations which must be reproduced in race evolution, as adaptations which the species must effect, are in these lines. The influence of organic selection is, therefore, exerted to determine, by the selection and accumulation of variations, the congenital equipment which most readily utilizes and supplements these traditional modes of behaviour. The two factors work together and for the same general result.

There is, therefore, in tradition a further determining factor. Natural selection plays about it to fix a requisite function here, to eradicate what is unnecessary and non-useful there — in short, by its omnipresent operation on this character and on that, to perfect the individual for the most adapted life.

It is here also that we touch upon the border line between psychophysical evolution and social evolution, a line which we may not now cross. Suffice it to say that once the community of tradition is established and the fitness of the individuals secured for a life in some degree of gregarious habit, and we then find the great bend in the line. Progress from now on ushers in the dominance of mind in the modes of conscious organization which characterize social life and institutions.

§ 3. *Intergenetic Concurrence*

The third question, mentioned above as involved in a full statement of the problem of determination, is that of the relation of the determination of development to that of evolution; that of the 'intergenetic' relation of the two lines of progress, growth, and descent.

We now find that the principles so far explained above, will, if they be true, afford an answer to this question also. The determination of the direction of evolution has been found to follow that of development. There is, therefore, in its great outline the 'concurrence' which the theory of recapitulation supposes, and which it is reasonable to expect if the correlations already[1] mentioned between the two series are actually realized. The determination of the individual's development is by a process of adjustment to a more or less stable environment. The evolution of the race is throughout, in its great features, a series of adaptations to the same bionomic conditions. Moreover, by the establishing of a tradition throughout the life history of the higher forms, there is set up a series of modes of behaviour to which, as we have seen, both development and evolution, by the operation of organic and natural selection, tend ever more approximately to conform. The two movements are, therefore, 'concurrent' in a very well-defined sense.

[1] Cf. Chap. XIII. § 5, below.

The recognition of the essentially psychophysical nature of the evolution process becomes increasingly imperative in the light of such a setting together of the subordinate problems in a single whole. We find as we advance a gradual shifting of the emphasis from the physical to the mental. This is not only true in respect to the sort of utilities which 'fit' variations subserve, but also in the very means of transmission itself. It is pointed out in the earlier work on *Social and Ethical Interpretations* that, as tradition advances, and with it a corresponding increase in the plasticity of the young who are educated in this tradition, social transmission comes directly to supersede the physical transmission of particular functions. Social transmission, however, is a process quite distinct from physical heredity. It has laws of its own.[1] The difference is so great that I have ventured to characterize social transmission as, in a sense, the means of the emancipation of mind from the limitations of biological progress; for by it there is secured a means of propagation of intelligent conduct without the negativing, swamping, and regressive effects of physical reproduction. Transmission by handing down, with imitative learning, is so different from transmission by physical heredity, that the series of conceptions which in the lower stages of evolution hold for both body and mind together — where both are subject to the single law of congenital variation with natural selection — are no longer common to them, but a series of additional conceptions emerge which are comparative principles principally in name. There are such differences in their operation in the two spheres respectively

---

[1] An attempt to work out certain of these laws is made in Chap. II. of the work just cited.

that instead of calling them comparative principles, we
may better denominate them 'analogies.'

§ 4. *Genetic Analogies*

Of the analogies drawn from organic evolution, which
spring up to vex the soul of the investigator in genetic
things in other fields, many are aspects of what is called
the 'biological analogy,' until now so much exploited in
the social sciences. Certain aspects of it are treated in the
papers which follow, and in the second volume of this
series referred to just above. For example, the 'struggle
for existence' is shown below (Chap. XV.) to take on three
quite different forms even in the animal world, where it
is a factor of direct importance in connection with the
operation of natural selection. In the same place, the facts
of conscious 'competition' and 'rivalry' are compared with
those of biological struggle, with the result that only under
certain conditions do they even show analogy with struggle for existence in the sense principally employed by
Darwin and Wallace — the struggle for food. So also,
when we come to subject the conception of 'selection' to a
thorough analysis, we have distinctions to make which for-
bid our using the biological conception in the mental and
social spheres except under the very restricted limitation,
namely, that the results of the selection in question normally fall under the laws of physical reproduction and
heredity for their conservation. Yet again, in the matter
of conservation of type, with regression, where there is the
question of the application of such a principle to mental
transmission, we find that the mental products do not, in
respect to their effectiveness for the future movement of

social evolution or development, follow such a law — that they follow, moreover, a very different and in no wise analogous law. The greater the variation in tradition, — the idea of the genius, the protest of a reformer, the new formulation of a scientific truth, — the greater may be its effect; while, by the law of biological regression, the great variation, the sport, tends to be swamped by interbreeding, and the wider his departure from the mean the less his chance of impressing his characters upon posterity. The whole case is summed up in the statement made above, to the effect that social progress is no longer under the limitations set by physical heredity; it is under the laws of mental process and organization.[1]

Some one may say, what is indeed quite true,[2] that this progress is after all due to the operation of natural selection, whereby the necessary plasticity required for the mind was selected and fixed; but such a statement alone would be quite inadequate as an explanation. For when so much is said, what is gained? So far may we go in the interpretation from the side of the physical; but the meaning, I submit, of evolution in this direction is not to be found on the side of plasticity but on the side of mind — the accommodations which are effected on the basis of the plasticity. We now, in short, recognize that wonderful endowment which is correlated with plasticity in the psychophysical whole. The emphasis in the interpretation of the twofold fact is not upon the process of the physical, but upon the events which are taking place in

[1] See the remarks on history, and especially the criticism of Professor Karl Pearson, in Chap. XIX. § 7.

[2] Professor Osborn, however, one of the original advocates of organic selection, does not admit that plasticity has been acquired through natural selection ; see the *American Naturalist*, Nov. 1897, cited in Appendix A.

the other aspect of the joint series. Hence we must draw directly upon the resources of that science, psychology, which makes the interpretation of the psychological movement its business.

§ 5. *Preformism and Accommodation*

There is here what seems to me to be a fundamental error in the general theory of preformism; and I shall state the point in a form in which it answers also a criticism of organic selection. It is said that the accommodations and modifications which are effected by the individual organism simply show the unfolding of what the congenital endowment of the creature has made possible; consequently, that these accommodations are sufficiently accounted for by the natural selection of the congenital variations which contribute to this endowment, so that there is nothing really additional or new in a theory which emphasizes these modifications.[1] That this is a partial truth only it is easy to show. It becomes evident so soon as we come to see that the characters which the individual develops are a compound, as has been said above, of his hereditary impulses with the forces of his environment. If it were simply a matter of continued reproduction without determinate evolution from generation to generation, then it would make no difference what the individuals might undergo during their lives, provided the germ-plasm remained unaffected. But so soon as it becomes a question of descent with adaptations which are selected from a great many possible ones, in intergenetic correlation with the modifications of individuals, then the question as to which

[1] Cf. the remarks on the relation of organic to natural selection in Chap. VIII. § 7.

variations are to be perpetuated and accumulated can be answered only by undertaking an investigation of this correlation; that is, by interpreting the actual accommodations, intelligent and other, which the individuals make. The use made of the plasticity by the intelligence, therefore, becomes the critically important thing for evolution theory, even though it assumes the presence of the plasticity itself.

It may be said, indeed, quite truly, that this value of accommodation is implicit in the theory of natural selection; for, according to that theory, there is continued selection of certain fit individuals, and their fitness may consist in their being plastic or 'accommodating.' This is especially true of the theory of Roux, which makes use of what he calls 'the struggle of the parts,' and of Weismann's 'intra-selection' theory. Yet still the qualifications of the fit individuals are not *given in their plasticity*, but they arise *only in the course of development;* and they may take on many different forms. There may be alternative ways in which the same plastic material or organism may adjust itself to the conditions of life. The same emergency may lead animals of common heredity and equal plasticity to make vital adjustments so different in kind that each may start a new line of evolutionary progress. In fact, I think many cases of divergent evolution have actually begun in such a situation (cf. Chap. XIII. § 2, 3). How, then, is it possible to say that both these differing lines of descent are equally accounted for by the same degree of plasticity in the individuals who are their common progenitors?

Suppose two creatures born with the same degree of plasticity in respect to a certain function, but with different correlations, or with differences in other characters which make their behaviour in effecting accommodations to the environment somewhat different. They adopt different ways of using their plastic substance and both live, yet with considerable differences of habit and behaviour. These, if there be enough individuals of each sort, would carry on from generation to generation their respective habits of life; tradition would spring up to set and confirm each group in its own way of life. And again there would be divergent or polytypic evolution as the result, although their original plasticity was the same. Here it is a question of the correlations of the plasticity, not merely the possession of it. In this case and the one just cited the *actual development dominates evolution*, not merely the possible development.[1] I am not able, therefore, to see great force in the contention of the preformists when they claim that the recognition of the variation by which a function is made possible in development supplies a sufficient theory of the course of the development, and also of its results in determining evolution.

All this is notably true in the matter of mind, and in evolution into which a strain of conscious accommodation has entered. Let us say, for instance, that the female bird has a certain capacity for preferential choice among possible males. This means nothing, unless she actually makes a choice. Then the physical characters of the offspring vary according as this male or that is chosen, and these go down to posterity. It is the *result which is the evolution*, and it is conditioned upon the use made of the endow-

[1] As Professor Poulton says, speaking of organic selection in general (see Appendix A), 'in this way natural selection would be compelled to act along a certain path' — a strong and true statement.

ment. We might as well say that a man is the cause of
all the follies of his wayward son because he begot the son,
as to say that natural selection is responsible for — or is
an adequate explanation of — the results which spring
from the accommodations of an organism, simply on the
ground that the plasticity of the organism has survived
by natural selection. Or, to take a case which more truly
depicts the function of natural selection, we might as
well say that the mother of Moses and the daughter of
Pharaoh were the essential factors in the production of that
great lawgiver's work, inasmuch as they warded off the
dangers which threatened his life.[1] But the endowments
of Moses would have been quite ineffective, despite his sal-
vation by the women, had not opportunities arisen for him
to use his gifts. His actual performance is what counted
in history; and so it is with the humblest organism which
accommodates itself to the environment, in so far as it
makes effective contribution to the characters of the generations which follow after it.[2]

[1] Yet even this figure is allowing too much to natural selection, for the
mother of Moses and the daughter of Pharaoh are, when considered as positive agents, more analogous to the positive accommodations which fit the
organism to survive ; it is these latter which save the creature's life. This
case may suffice to show how impossible it is to put one's finger on anything positive to represent natural selection. Of course all will admit that the
recognition of the actual facts and factors is the main thing — not the naming
of them. Yet questions of the relative rôles of the factors are important, both
for interpretation and for the integrity of our logic.

[2] Professor James Ward, art. 'Psychology,' in the *Ency. Brit.*, 9th. ed., was
one of the earlier writers who pointed out that organisms act very positively in
adjusting themselves to their environment, selecting and even changing their
life conditions by their own acts. He called this 'subjective selection,' and
he has developed in a later publication, *Naturalism and Agnosticism*, the possible influence this might be expected to have on the future course of
evolution, uniting with it, however, the theory of the inheritance of acquired
characters.

Returning to our main subject, after this digression, we may emphasize the necessity, now so often pointed out, of taking up, wherever possible, the psychophysical point of view, and of recognizing, as of equal importance with the biological, those factors and processes which, it may be, the psychologist alone is able to describe. No better instance can be cited — in illustration of many of the considerations so far advanced — than the problem of the origin of instinct, of which certain phases are treated in the following pages.

E

PART II

THE METHOD OF EVOLUTION

CHAPTER IV

THE PLACE OF CONSCIOUSNESS IN EVOLUTION [1]

§ 1. *Professor Cope's Table*

In a table in the *Monist*, July 26, Professor Cope gives certain positions on points of evolution, in two contrasted columns, as he conceives them to be held by the two groups of naturalists divided in regard to heredity into Preformists and the advocates of Epigenesis. The peculiarity of the Epigenesis column is that it includes certain positions regarding consciousness, while the Preformist [2] column has nothing to say about consciousness. Being struck with this I wrote to Professor Cope — the more because the position ascribed to consciousness seemed to be the same, in the main, as that which the present writer has developed, from a psychological point of view, in the work on *Mental Development*. I

---

[1] From *Science*, Aug. 23, 1895 (an informal communication).

[2] Preformism is the view of those who hold that the individual organism is 'preformed' in the germ and its development is in some way an unfolding of preformed parts. Epigenesis holds to a real growth or production of parts in the developing organism. Professor Cope holds with many Epigenesists that these newly acquired parts or functions are inherited ('Lamarckian' heredity and evolution); and by the term 'Preformism' he designates the opposed (Darwinian) view of heredity and evolution. The terms Lamarckism and Darwinism are used in the following pages to express this contrast.

learn from him that the table (given herewith) is not new; but was published in the 'annual volume of the Brooklyn Ethical Society in 1891': and the view which it embodies is given in the chapter on 'Consciousness in Evolution,' in his work, *The Origin of the Fittest* (1887).

1. Variations appear in definite directions.	1. Variations are promiscuous or multifarious.
2. Variations are caused by the interaction of the organic being and its environment.	2. Variations are 'congenital,' or are caused by mingling of male and female germ-plasms.
3. Acquired variations may be inherited.	3. Acquired variations cannot be inherited.
4. Variations survive directly as they are adapted to changing environments (natural selection).	4. Variations survive directly as they are adapted to changing environments (natural selection).
5. Movements of the organism are caused or directed by sensation and other conscious states.	5. Movements of the organism are not caused by sensation or conscious states, but are a survival through natural selection from multifarious movements.
6. Habitual movements are derived from conscious experience.	6. Habitual movements are produced by natural selection.
7. The rational mind is developed by experience through memory and classification.	7. The rational mind is developed through natural selection from multifarious mental activities.

Apart from the question of novelty in Professor Cope's positions — and that one should have supposed them so can only show that one had read hastily, not having earlier become acquainted with Professor Cope's views — I wish to point out that the placing of consciousness, as a factor in the evolution process, exclusively in the Lamarckian column, appears quite unjustified. It is not a question of a causal interchange between body and mind. It is not likely that any naturalist would hold to an injection of energy in any form into the natural processes, on the part of consciousness; though, of course, Professor Cope himself can say whether such a construction is true in his case.[1] Many psychologists are about done with a view like that. The question at issue when we ask whether consciousness has had a part in the evolutionary process is really that as to whether we may say that the presence of consciousness — in the form say of sensations of pleasure and pain — with its correlative nervous or organic processes, has been an essential factor in evolution; and if so, further, whether its importance is because it is in alliance with the consciousness aspect that the organic aspect gets in its work. Or, to take a higher form of consciousness — does the memory of an object as having given pleasure modify the organism's reaction to that object the second time? Such may be the case, even though it is only the physical basis of memory that has an efficient causal relation to the other organic processes of the animal.

Conceiving of the function of consciousness, therefore, as in any case not that of a *deus ex machina*, the question at issue is whether it can have an essential place in the evolution process as the Darwinians construe that process. Professor Cope believes not.[2] I believe that the place of consciousness may be the same — and may be the essential place that Cope gives it in his left-hand column and which is given to it in *Mental Development* — on the Darwinian view. I have argued briefly for this indifference to the particular theory one holds of heredity, in the volume referred to,[3] reserving for later pages certain arguments in detail based upon the theory of the

---

[1] In a reply made to this paper by Professor Cope he declares for such a view (*American Naturalist*, April, 1896, p. 342); see the next Chapter.

[2] See his *Primary Factors of Organic Evolution*.   [3] Chap. VII.

individual's personal relation to his social environment. The main point involved, however, may be briefly suggested here, although, for the details of the influences now indicated, the other book may be again referred to (chapters on 'Suggestion' and 'Emotion').[1]

The writer there traces in some detail what other writers also have lately set in evidence, *i.e.*, that in the child's personal development, his ontogenesis, his life history, he works out a faithful reproduction of his social conditions. He is, from childhood up, excessively receptive to social suggestion; his entire learning is a process of conforming to social patterns. The essential to this, in his heredity, is very great plasticity, cerebral balance and equilibrium, a readiness to overflow into the new channels which his social environment dictates. He has to learn everything for himself, and in order to do this he must begin in a state of great plasticity and mobility. Now, my point, put briefly, is that these social lessons which he learns for himself take the place largely of the heredity of particular paternal acquisitions. The father must have been plastic to learn, and this plasticity is, so far as the evidence goes, the nervous condition of consciousness; thus the father learned, through his consciousness, from his social environment. The child does the same. What he inherits is the nervous plasticity and the consciousness. He learns particular acts for himself; and what he learns is, in its main lines, what his father learned. So he is just as well off, the child of Darwinism, as if he were physical heir to the acquisitions which his father made. This process has been called 'Social Heredity,' seeing that the child really comes into possession of the details;

[1] Also the later work, *Social and Ethical Interpretations.*

but he comes by them socially, through this process of social growth, rather than by direct physical inheritance.

To show this in a sketchy way, we may take the last three points which Professor Cope places under the Lamarckian column, the points which involve consciousness, and show how indeed they may still be true for the Darwinian if he avail himself of the resource offered by 'Social Transmission.'

This is done rather from interest in the subject than with any wish to controvert Professor Cope ; and it may well be that his later statements may show that he is able to accept the argument.[1]

§ 2. *The Origin of Adaptive Movements*

1. (5 of Cope's table.) 'Movements of the organism are caused or directed by sensation and other conscious states.'

The point at issue here between the advocates of the two views of evolution would be whether it is necessary that the child should inherit any of the particular conscious states, or their special nervous dispositions, which the parent acquired in his lifetime, in order to secure through them the performance of the same actions by the child. I should say, no ; and for the reason — additional to the usual arguments of the Darwinians — that 'Social Transmission' is sufficient to secure the result. All we have to find in the child is the high consciousness represented by the tendency to imitate socially and so to absorb social copies, together with the law widely recognized by psychologists under the name of dynamogenesis — *i.e.*, that the thought

[1] In his reply, referred to above, Professor Cope fully accepts the fact called here 'Social Heredity.'

of a movement tends to discharge motor energy into the channels as near as may be to those necessary for that movement.[1] Given these two elements of endowment in the child, and he can learn anything that his father did, without inheriting any particular acts learned by the parent. And we must in any case give the child so much ; for the principle of dynamogenesis is a fundamental law in all organisms, and the tendency to learn by imitation, suggestion, etc., is present, as a matter of fact, with greater or less range, in man and in many other animals as well.

The only apparent hindrance to the child's learning everything that his life in society requires would be just the thing that the advocates of Lamarckism argue for — the inheritance of acquired characters. For such inheritance would tend so to bind up the child's nervous substance in fixed forms that he would have less or possibly no plastic substance left to learn anything with. Such fixity occurs in the animals in which instinct is largely developed ; they have little power to learn anything new, just because their nervous systems are not in the mobile condition represented by high consciousness. They have instinct and little else. Now, I think the Darwinian can account for instinct also, but that is beside the point; the point to be made now is that, if Lamarckism were true, we should all be, to the extent to which both parents perform the same acts (as, for example, speech) in the condition of the creatures who do only certain things and do them by instinct. It may well be asked of the Lamarckian: What is it that is peculiar about the strain of heredity of certain creatures that they should be so remarkably endowed with instincts ?

[1] Both of these requirements are worked out in detail in *Mental Development.*

Must he not say in substance that the nervous material of these creatures has been 'set' in the creatures' ancestors? Then why are not all constant functions thus set? But the question of instinct is touched upon under the next point.

2. (6 of Cope's table.) 'Habitual movements are derived from conscious experience.' This may mean movements habitual to the individual or to the species in question. If it refers to the individual it may be true on either doctrine, provided we once get the child started on the movement — a point discussed in other connections.[1] If, on the other hand, habitual movements mean movements characteristic of species, we raise the question of race habits, best typified in instinct. Agreeing that many race habits arose as conscious functions in the first place, and making that our supposition, again we ask: Can one who believes it still be a Darwinian? It would appear that he could. The problem set to the Darwinian would not in this case differ from that which he has to solve in accounting for evolution generally; it would not be altered by the postulate that consciousness is present in the individual. He may say that consciousness is a variation, and what the individual does by it follows from this variation. And then what later generations do through their consciousness is all given with the variations which they constitute on the earlier variations.[2] In other words, I do not see that the case is made any harder for the Darwinian by the postulate that consciousness with its nervous correlate is a real factor.

[1] Chap. VII. § 3; Chap. VIII. § 6; Chap. IX. § 2; Chap. XVII. § 5.

[2] This is said by thorough-going preformists (Weismannists). But I think this case is much simplified by the hypothesis of 'organic selection' (developed in the following papers) of which the following paragraphs are a summary statement (notably the lines now italicized).

Indeed, we may well go still further and say that the case is easier for him when we take into account the phenomenon of social heredity. In children, for example, there are great variations in mobility, plasticity, etc. — in short, in the ease of operation of social heredity as seen in the acquisition of particular functions. Children are notoriously different in their aptitudes for acquiring speech, for example; some learn faster, better, and more. Let us say that this is true in animal companies generally; then *the most plastic individuals will be preserved to do the advantageous things for which their variations show them to be the most fit. And the next generation will show an emphasis of just this direction in its variations.* So the fact of social acquisition — the fact of acute use of consciousness in ontogeny — becomes an element in phylogeny, also, even on the Darwinian theory.

Besides, when we remember that the permanence of a habit learned by one individual is largely conditioned by the learning of the same habits by others (notably of the opposite sex) in the same environment, we see that an enormous premium must have been put on variations of a social kind — those which brought different individuals into some kind of joint action or coöperation. *Wherever this appeared, not only would habits be maintained, but new variations, having all the force of double hereditary tendency, might also be expected.* But consciousness is, of course, the prime variation through which coöperation is secured. All of which means, if it be true, that the rise of consciousness is of direct help to the Darwinian in accounting for race habits — notably those which are in some degree gregarious, coöperative, or social.

3. (7 of Cope's table.) 'The rational mind is developed

by experience, through memory and classification.' This, too, is true, provided the term 'classification' has a meaning that psychologists agree to. So the question is again: Can the higher mental functions be evolved from the lower without calling in use-inheritance? So it seems. Here indeed it seems that the fact of social transmission is the main and controlling consideration. It is notorious how meagre the evidence is that a son inherits or has the peculiar mental traits of parents beyond those traits contained in the parents' own heredity. Galton has shown how rare a thing it is for artistic, literary, or other marked talent to maintain its strength in later generations. Instead, we find such endowments showing themselves in many individuals at about the same time, in the same communities, and under common social conditions. Groups of artists, musicians, literary men, appear together — as it were, a social outburst. The presuppositions of genius — obscure as the subject is — seem to be great power of learning or absorbing, marked gifts or proclivities of a personal kind which are not present in the parents but fall under the head of variations ; and with these a social environment of high level in the direction of these variations. The details of the individual's development, inside of the general proclivity which he has, are determined by his social environment, not by his natural heredity. And no doubt the phylogenetic origin of the higher mental functions — thought, self-consciousness, etc. — must have been similar.[1]

There is not any great amount of truth in the claim of Spencer that intellectual progress in the race requires the Lamarckian view. The level of culture in a community

---

[1] Detailed account of the social factors involved in the evolution of these higher faculties is attempted in the two earlier volumes of this series.

seems to be about as fixed a thing as moral qualities are capable of being, much more so than the level of individual endowment. This latter seems to be capricious or variable, while the former proceeds by a regular movement and with a massive front. It would seem, therefore, that intellectual and moral progress is gradual improvement, through improved relationships on the part of the individuals to one· another ; a matter of social accommodation, rather than of direct natural inheritance on the part of individuals. It is only a rare individual whose heredity enables him to break through the lines of social tissue and imprint his personality upon the social movement. And in that case the only explanation of him is that he is a variation, not that he inherited his intellectual or moral power. Furthermore, I think the actual growth of the individual in intellectual stature and moral attainment can be traced in the main to certain of the elements of his social *milieu*, allowing always a balance of variation in the direction in which he finally excels.

So strong does the case seem for the social heredity view in this matter of intellectual and moral progress that I may suggest an hypothesis which may not stand in court, but which seems interesting. May not the rise of the social life be justified from the point of view of a second utility in addition to that of its utility in the struggle for existence as ordinarily understood, the second utility, *i.e.*, of giving to each generation the attainments of the past which physical heredity is inadequate to transmit ? Whether we admit Lamarckism or confine ourselves to Darwinism, I suppose we may safely accept Galton's law of Regression and Weismann's principle of Panmixia in some form. Now as social life advances we find the beginning of the

artificial selection of the unfit ; and so these negative principles begin to work directly in the teeth of progress, 'as
many writers on social themes have recently made clear.
This being the case, some other resource is necessary besides physical heredity. On my hypothesis it is found in
the common social standards of attainment to which the
individual is fitted to conform and to which he is compelled to submit. This secures progress in two ways :
First, *by making the individual learn what the race has
learned,* thus preventing social retrogression, in any case ;
and second, *by putting a direct premium on variations which
are socially available.*

Under this general conception we may bring the biological phenomena of infancy, with all their evolutionary
significance : the great plasticity of the mammal infant as
opposed to the highly developed instinctive equipment of
other young ; the maternal care, instruction, and example
during the period of helplessness ; and the very gradual
attainment of the activities of self-maintenance in conditions in which social activities are prominent or essential.
All this stock of the evolution theory is available to confirm
this view.

And to finish where we began, all this is through that
wonderful instrument of acquisition, consciousness ; for
consciousness is the avenue of all social influences.

CHAPTER V

Heredity and Instinct[1]

§ 1. *Romanes on Instinct*

In his able posthumous work on *Post-Darwinian Questions, Heredity and Utility*, the lamented G. J. Romanes sums up the evidence for the inheritance of acquired characters in the final statement that only two valid arguments remain on the affirmative side; and to each of these arguments he has devoted considerable space. One of these arguments is from what he calls 'selective value,' and the other from the 'co-adaptations' found in the instincts of animals. He says (p. 141): 'Hence there remain only the arguments from selective value and co-adaptation.' If we take the instincts as illustrating the application of the principle of 'selective value as well,' we may gather the evidence which Romanes was disposed to cling to, for the inheritance of acquired characters, into a single net, and inquire as to the need of resorting to the Lamarckian factor in accounting for the origin of instinct. I wish to suggest some considerations from the psychological side, which seem to me entirely competent to remove the force of these two arguments, and to show to that extent that the instincts can be accounted for without appeal to the hypothesis of 'lapsed intelligence,' as the use-inheritance theory,

[1] Discussion (revised) following Professor C. Lloyd Morgan before the New York Academy of Sciences, Jan. 31, 1896; from *Science*, March 20, 1896.

in its application to this problem of instinct, is called; in other words, to show that Darwin, Romanes, and the Neo-Lamarckians are not right in considering instinct as 'inherited habit.'

§ 2. *Instinct and Lamarckism: Co-adaptation*

The argument from co-adaptation in the case of instinct requires the presence of some sort of intelligence in an animal species, the point being that since the coördination of muscular movements found in the instincts are so co-adapted they could not have arisen by gradual variation. Partial adaptations tending in the direction of an instinct would not have been useful; and intelligence alone would suffice to bring about the coördinations which are too complex to be accounted for as spontaneous variations. These intelligent coördinations then become habits by repetition in the individual and show themselves in later generations as inherited habits due to 'lapsed intelligence.' Assuming, then, with Romanes — whom we may cite as a very recent upholder of the view — the existence of some intelligence in a species antecedently to the appearance of the instinct in question, we may be allowed that supposition and resource.

I. But now let us ask how the intelligence brings about coördinations of muscular movement. The psychologist is obliged to reply: Only by a process of selection (through pleasure, pain, experience, association, etc.) from certain alternative complex movements which are already possible to the individual animal. These possible combinations are already there, born with him, or resulting from his previous habits. The intelligence can never, by any possibility, create a new movement; nor effect a new combination of movements, if the apparatus had not been made already trained by actual use for the combination which is effected.[1] So far as there are modifications in the grouping, even these are very slight functional variations from the uses already made of the muscles involved. This point is no longer subject to dispute; for pathological cases show that unless some adequate idea of a former movement made by the same muscles, or some other idea which stands for it by association, can be brought up in mind, the intelligence is helpless. Otherwise it cannot only not make new movements; it cannot even repeat old habitual movements. So we may say that intelligent adaptation does not create coördinations; it only makes functional use of coördinations which were alternatively present already in the creature's equipment.[2]

Interpreting this in terms of congenital variations, we may say that the variations which the intelligence uses are alternative possibilities of muscular movement. But these are exactly the variations which instinct uses, except that in instinct they are not alternative. That this is so, indeed, lies at the basis of the claim that instinct is inherited habit. The real difference in the variation involved in the two cases is in the connections in the brain whereby in

---

[1] Professor Cope has understood this to mean that consciousness can select out or direct the combination. This is accomplished, in my opinion, by a process analogous to natural selection, *i.e.*, the survival of useful movements from overproduced movements, a process called 'functional selection' in *Mental Development*, formulated in an earlier paper, 'The Origin of Volition,' reprinted in *Fragments in Philosophy and Science* (1902); see also the references given, p. 56, note 1, above.

[2] When we strain our muscles to accomplish a new act of skill, we are aiming to use the apparatus in new ways by a selection from possible combinations; and even when we learn to use disused muscles, as those of the ear, we are only aiming to stir up possible connections not before actively used.

instinct the muscular coördination is brought into play *directly* by a sense stimulation; while in intelligence it is brought into play *indirectly*, *i.e.*, through association of brain processes, with selection of fortunate combinations. Now this difference in the central brain connections is, I submit, not at all a great one, relatively speaking, and it might well be due to spontaneous variations. The point of view which holds that great co-adaptations of the muscles have to be acquired *all at once* by the creature is quite mistaken.

§ 3. *Instinct and Lamarckism:* '*Selective Value*'

The same class of considerations refutes the argument from 'selective value.'[1] This argument holds that the instinct could not have arisen by variations alone, with natural selection, since partial coördinations tending in the direction of the instinct would not have been useful; so the creatures with such partial coördinations merely would have been killed off, and the instinct could never have reached maturity; only variations which are of sufficient value or utility to be 'selective' would be kept alive and perfected.

But we see that the intelligence which is appealed to, to take the place of instinct and to give rise to it, uses just these partial variations which tend in the direction of the instinct; so the intelligence *supplements* such partial coördinations, makes them functional, and so *keeps the creature alive*. This prevents the 'incidence of natural selection,' to use a phrase of Professor Lloyd Morgan's. So the supposition that intelligence is operative turns out to be just

[1] In my opinion 'selective value' is equivalent simply to 'utility': any amount of utility is 'selective.'

the supposition which makes the use-hypothesis unnecessary. *Thus kept alive, the species has all the time necessary to perfect the variations required by a complete instinct.*[1] And when we bear in mind that the variation required is, as was shown above, not on the muscular side to any great extent, but in the central brain connections, and is a slight variation for functional purposes at the best, the hypothesis of use-inheritance becomes, to my mind, not only unnecessary, but quite superfluous.

§ 4. *Social Transmission and Instinct*

II. There is also another great resource open to the Darwinian in this matter of instinct; also a psychological resource. Weismann and others have shown that the influence of animal intercourse, seen in maternal instruction, imitation, gregarious coöperation, etc., is very important. Wallace dwells upon the actual facts which illustrate the 'imitative factor,' as we may call it, in the personal development of young animals. It is argued above that Spencer and others are in error in holding that social progress demands the use-inheritance hypothesis;[2] since the socially-acquired actions of a species, notably man, are socially handed down, giving a sort of 'social transmission' which supplements physical heredity. And when we come to inquire into the actual mechanism of imitation on the part of a young animal, we find much the same sort of function involved as in intelligent adaptation. The impulse to imitate requires the ability to act out for him-

---

[1] Italicized in this reprinting (as is done in the preceding paper) as anticipating the full statement of the theory of 'Organic Selection' later on.

[2] Cf. *Science*, Aug. 23, 1895, the preceding paper; summarized in *Nature*, Vol. LII., 1895, p. 627.

F

self certain of the actions which the animal sees, to make the sounds which he hears, etc. Now this involves connections of the centres of sight, hearing, etc., with certain muscular coördinations. If he have not the coördinations, he cannot imitate; just as we saw above is the case with intelligence, if the creature have not the function ready, he cannot perform it intelligently. Imitation differs from intelligence in being a general form of coördinated adaptation, while intelligence involves a series of special forms.[1] But both make use of the apparatus of coördinated movement. So we find, as an actual fact generally agreed upon, that by imitation the little animal picks up directly the example, instruction, mode of life, etc., of his private family circle and of his species.[2] This, then, enables him to use effectively, for the purposes of his life, the coördinations which become instincts later on in the life of the species; and again we have here two points which directly tend to neutralize the arguments of Romanes from 'selective value' and 'co-adaptation.' The co-adaptations may be held to be gradually acquired, since the coördinations of a partial kind are utilized by the imitative functions before they become instinctive. And the law of 'selective value' does not get application, since the imitative functions, by using these muscular coördinations, *supplement them, secure accommodations, keep the creature alive, prevent the 'incidence of natural selection,' and so give the species all the time necessary to get the variations required for the full instinctive performance of the function.*

---

[1] That they are really the same in type and *origin* is argued in detail in the work *Mental Development.*

[2] Largely along the line of his native impulses, as recent researches have shown (1902).

III. These positions are illustrated in a very fortunate way by the interesting cases reported by Professor Ll. Morgan in his instructive discussion. He cites the beautiful observation that his young chicks had the instinct to drink by throwing their heads up in the air, etc., but that it came into action only after they had the taste[1] of water by accident or by imitating the old fowl. As Ll. Morgan says, the 'incidence of natural selection' is prevented by imitation or instruction or intelligent adaptation (in cases where experience is required). So, in this instance, the instinct of drinking, which only goes so far as a connection of certain muscular coördinations with the sense of taste (wet bill) is made effective for the life interests of the chick. *Thus kept alive the species has plenty of time — in case it should be necessary — to get a connection established* also between the sight centre and the same coördination of movements ; so that future chicks may be born with a capacity for drinking when water is seen only, without waiting for instruction, a fortunate accident, or an example to imitate. So we may imagine creatures, whose hands were used for holding on with the thumb and fingers on the same side of the object held, to have first discovered, under stress of circumstances and with variations which permitted the further adaptation, how to make intelligent use of the thumb for grasping opposite to the fingers, as we now do. Then, let us suppose that this proved of such utility that all the young that did not do it were killed off ; the next generation following would be intelligent or imitative enough to do it also. They would

[1] Or other form of stimulation from getting the bill wet (this in view of a later discussion, as to just what the stimulation is, in *Science*) — reprinted by Mills in *Nature and Development of Animal Intelligence*, pp. 277 ff.

use the same coördinations intelligently or imitatively, prevent natural selection getting into operation, *and so instinctive 'thumb-grasping' might be waited for indefinitely by the species and then arise by accumulated variation, altogether apart from use-inheritance.*

We may say, therefore, that there are two great kinds of influence, each in a sense hereditary : there is *physical heredity* by which variations are congenitally transmitted with original endowment, and there is ' *social heredity* ' by which functions socially acquired (*i.e.*, imitatively, covering all the conscious acquisitions made through intercourse with other animals) are socially transmitted. The one is phylogenetic ; the other, ontogenetic. But these two lines of transmission are not separate nor are they uninfluential on each other. Congenital variations, on the one hand, are kept alive and made effective by their conscious use for intelligent and imitative accommodations in the life of the individual ; and, on the other hand, intelligent and imitative accommodations become congenital *by further progress and refinement of variation in the same lines of function as those which their acquisition by the individual called into play.* But there is no need in either case to assume the Lamarckian factor.

The intelligence holds a remarkable place in each of these categories. It is itself, as we have seen, a congenital variation ; but it is also the great agent of the individual's personal accommodations both to the physical and to the social environment.

The emphasis, however, of the first of these two lines of transmission gives prominence to instinct in animal species, and that of the other to the intelligent and social coöperation which goes on to be human. The former represents a tendency to brain variation in the direction of fixed connections between certain ,sense-centres and certain groups of coördinated muscles. This tendency is embodied in the white matter and the lower brain centres. The other represents a tendency to variation in the direction of alternative possibilities of connection of the brain centres with the same or similar coördinated muscular groups. This tendency is embodied in the cortex of the hemispheres. I have cited 'thumb-grasping' because we may see in the child the anticipation, by intelligence and imitation, of the use of the thumb for the adaptation which the simian probably gets by instinct or accident, and which I think an isolated and weak-minded child, say, would also come to acquire by instinct or accident when his apparatus became sufficiently matured.

§ 5. *Instinct and Intelligence*

IV. Finally there are two general bearings of the position taken above regarding the place and function of intelligence and imitation which may be briefly noted.

1. We reach a point of view which gives to organic evolution a sort of intelligent direction after all; for of all the variations tending in the direction of an instinct, but inadequate to its complete performance, *only those will be supplemented and kept alive which the intelligence ratifies and uses for the animal's individual accommodations.* The principle of selection applies strictly to the others or to some of them. So natural selection eliminates the others; and the *future development of instinct must at each stage of a species' evolution be in the directions thus ratified by intelligence.* So also with imitation. Only those imitative

actions of a creature which are useful to him will survive in the species; for in so far as he imitates actions which are injurious, he will aid natural selection in killing himself off. So intelligence, and the imitation which copies it, will set the direction of the development of the complex instincts even on the Darwinian theory; and in this sense we may say that consciousness is a 'factor' without resorting to the vague postulates of 'self-adaptation,' 'growth-force,' 'will-effort,' etc., which have become so common of late among the advocates of the new vitalism.

2. The same consideration may give the reason in part that instincts are so often coterminous with the limits of species. Similar creatures find similar uses for their intelligence, and they also find the same imitative actions to be to their advantage. So the interaction of these conscious factors with natural selection brings it about that the structural definition which characterizes species, and the functional definition which characterizes instinct, largely keep to the same lines.

CHAPTER VI

HEREDITY AND INSTINCT (II.)[1]

In the preceding chapter I argued from certain psychological truths for the position that two general principles recently urged by Romanes for the Lamarckian, or 'inherited habit,' view of the origin of instincts do not really support that doctrine. These two principles are those cited by Romanes under the phrases respectively 'co-adaptation' and 'selective value.' In the case of complex instincts these two arguments really amount to but one, so long as we are talking about the *origin* of instinct. And the one argument is this: that partial co-adaptations in the direction of an instinct are not of selective value; hence instinct could not have arisen by gradual partial co-adaptive variations, but must have been acquired by intelligence and then inherited. This general position is dealt with in the earlier chapter.

It will be remembered, however, that the force of the refutation of the Lamarckian's argument on this point depends on the assumption, made in common with him, that some degree of intelligence or imitative faculty is present before the completion of the instinct in question. To deny this is, of course, to deny the contention that instinct is 'lapsed intelligence,' or 'inherited habit.' To assume it, however, opens the way for certain further questions, which I may now take up briefly, citing Romanes by preference as before.

[1] Conclusion of the preceding paper, printed separately in *Science*, April 10, 1896.

§ 1. *Duplicated Functions*

I. The argument from 'selective value' has a further
and very interesting application by Romanes. He uses
the very fact upon which the argument in the earlier
pages is based to get further support for the inheritance
of habits. The fact is this, that intelligence may perform
the *same acts* that instinct does. So granting, he argues,
that the intelligent performance of these acts comes first
in the species' history, this intelligent performance of the
actions serves all the purposes of utility which are claimed
for the instinctive doing of the same actions. If this be
true, then variations which would secure the instinctive
performance of these actions do not have selective value,
and so the species would not acquire them by the operation of natural selection. By the Lamarckian theory, however, he concludes, the habits of intelligent action give
rise to instincts for the performance of the same actions
which are already intelligently performed, the duplicate
functions often existing side by side in the same creature.[1]

This is an ingenious turn, and raises new questions of
fact. Several things come to mind in the way of comment.

First. It rests evidently on the state of things required
by my earlier argument against the Lamarckian claim
that co-adaptation could not have been gradually acquired
by variation; the state of things which shows the intelligence preventing the 'incidence of natural selection' by
supplementing partial co-adaptation. Romanes now assumes that intelligence prevents the operation of natural
selection on further variations, and so rules out the origin

---

[1] *Op. cit.*, pp. 74–81.

of instinct through that agency; or, put differently, that
actions which are of selective value when performed intelligently are not afterwards of selective value when performed
instinctively. But this seems in a measure to contradict
the argument which is based on co-adaptations (examined
in the earlier pages), *i.e.*, that instincts could not have
arisen by way of partial co-adaptations at all. In other
words, the argument from ' co-adaptations ' asserts that the
partial co-adaptations are not preserved, being useless; that
from selective value asserts that they are preserved and,
with the intelligence thrown in, are so useful as to be of
selective value. We have seen that the latter position is
probably the true one; but that the inheritance of acquired
characters is then, through this union of variation with
intelligence, made unnecessary.

Second. Assuming the existence side by side in the
same creature of the ability to do intelligently certain
things that he also does instinctively, it is extraordinary
that Romanes should then say that the instinctive reflexes
have no utility additional to that of the intelligent performance. On the contrary, the two sorts of performance
of the same action are of very different and each of extreme
utility. Reflex actions are quicker, more direct, less
variable, less subject to inhibition, more deep-seated organically, and thus less liable to derangement. Intelligent
actions—the same actions in kind—are, besides the points
of opposition indicated, and by reason of them, more
adaptable. Then there is the remarkable difference that
intelligent actions are centrally stimulated, while reflex
actions are peripherally stimulated. We cannot go into all
these differences here; but the case may be made strong
enough by citing certain divergencies between the two

sorts of performance, with illustrations which show their separate utilities.

1. Reflex and instinctive actions are less subject to derangement. Emotion, shock, temporary ailment, hesitation, aboulia, lack of information, etc., may paralyze the intelligence; and instinct and reflex action may keep the creature alive in the meantime. What keeps dogs alive, and able to meet the demands made upon them, after extended ablation of the brain cortex?

2. Reflexes are quicker. Suppose instead of winking reflexly when a foreign body approaches the eye, I waited to see whether it was near enough to be dangerous, or even shut my eye as quickly as I could; I should join the ranks of the blind in short order.

3. Reflex actions are more deep-seated, and arose genetically first. What keeps the infant alive and in touch with his environment before the voluntary fibres are developed? This genetic utility alone would seem critical enough to justify most of the genuine reflexes of the organism, — supplemented, of course, in the human case, by the mother!

4. Intelligent actions are centrally stimulated. This means that brain processes release the energy which goes out in movement, and that something earlier must stimulate the brain processes. This something is association in some shape between present stimulating agencies in the environment and memories with pleasures or pains. In other words, certain central processes intervene between the outside stimulus and the release of the energies of movement. In reflexes, however, no such central influence intervenes. The stimulus in the environment passes directly — is reflected — into the motor apparatus. Hence

the reflex is more direct, undeviating, invariable, sure. For example, research has recently proved that involuntary movements may be produced in a variety of normal circumstances, and in hysterical subjects, when the stimulation is too weak, or intermittent, or unimportant, to be perceived at all.

5. Experiments show that the energies of the two are not quantitatively the same. Mosso and Waller have shown that the muscles do work under electric stimulation after being quite exhausted for voluntary action, and *vice versa*. There may be exchanges of energy between the two circuits involved, and this may give the animal increased force in this reaction or that.

6. The intelligence could not attend to the necessary functions of life without the aid of reflexes — to say nothing of the luxuries of acquisition. So not to have the reflexes would prevent the growth of the intelligence. For example, suppose we had to walk, wink, breathe, swallow, brush away flies and mosquitoes, etc., all by voluntary attention to the details and all at the same time. While chasing flies we should forget to breathe! And when should we have a moment's time to think? In this line it is in order to cite the experiments made on 'distraction,' which show that most of the common adaptations of life can go on by reflex and subconscious processes while the intelligence is otherwise occupied.[1]

7. Attention and voluntary intermeddling with reflex and instinctive functions tend to destroy their efficiency, bringing confusion and all kinds of disturbance.

The foregoing are all psychological facts, and more might be added showing that instinct has its own great

[1] See Binet, *Alterations of Personality*, Part II., Chap. V. (Eng. trans.).

utility even when the intelligence may perform the same actions in its own fashion. So it remains in each case to find out this utility and appraise it, before we say that it is not a reason for survival. It would seem that reflexes are of supreme importance and value; and if so, then natural selection may be appealed to, to account for them. So about all that remains of this argument of Romanes is the contribution which it makes to the refutation of his other one — from co-adaptations. The assumption of intelligence disposes of both the arguments, for the intelligence supplements slight co-adaptations and so makes them effective and useful; but it does not keep them from serving other utilities, as instincts, reflexes, etc., by further variation.

§ 2. *Reflexes and Imitation*

II. There is still another very interesting question also to be settled by fact. Romanes and others cite simple reflexes as well as complex instincts as giving illustrations of the application of the principle of 'inherited habit' or 'lapsed intelligence'; and the cases which Romanes lays great stress on are the reflex actions of man's withdrawal of the leg from irritation to the soles, and the brainless frog's balancing himself.[1] The Neo-Lamarckian theory requires the assumption of intelligence for all of these. I have shown that granting the intelligence, that is just the assumption which in many cases enables us to discard the Lamarckian factor. But we may ask: Is the intelligence necessary for all reflexes?

The question is too involved for treatment here; but

---

[1] Passage cited above from Romanes.

the assumption that intelligence is necessary in any sense which makes the *conscious voluntary* performance of the action always precede the reflex performance in evolution is difficult to defend. For all that we know of the brain seat of voluntary intelligence, of the use of means to ends, etc., indicates that such action is dependent upon the presence of the great mass of organic reflex processes which go on below the cortex. Complex associative processes must be genetically (and phylogenetically) later than the simple reflex processes, which, as has been intimated above, they presuppose.

But the more liberal definition of intelligence, which makes it include all kinds of conscious processes—the assumption of intelligence being that simply of conscious process of some kind—that is a different matter. This supposition seems to be necessary on either theory of instinct, as is argued above; for if we do not assume it, then natural selection is inadequate, as say Romanes and Cope; but if we do assume it, then the inheritance of acquired characters is unnecessary. On this simpler definition of intelligence, however, we find certain states of consciousness, of which imitation is the most prominent example, serving nature a turn in the matter of evolution.

On this wider definition of intelligence the difference between intelligent (*e.g.*, imitative) action and instinctive reflex action is much greater than that pointed out in detail above between voluntary and reflex action. A word to show this may be allowed here, since it makes yet stronger the case against the special argument from selective fitness, which this paper set out to examine.

The differences between imitative action and reflex or

instinctive action are not just those which we have found between voluntary and reflex actions. Imitation seems to be a native impulse; and in so far it seems to be, like the instincts, stimulated from the periphery. But it has a further point of differentiation from the special instincts and reflexes in that it is what has been called a 'circular' reaction, *i.e.*, it tends to reproduce its stimulus again, — the movement seen, the sound heard, etc. There is always a certain comparability or similarity, in a case of conscious imitation, between the thing imitated and the imitator's result; and the imitation is unmistakably real in proportion as this similarity is real. We may say, therefore, that consciously imitative actions are confined to those certain channels of discharge with produce results comparable with the 'copy' which is imitated.

But the special instincts and reflexes are not so. They show the greatest variety of arrangement between the stimulus and the movement which results from it — arrangements which have grown up under the law of survival. They represent, therefore, special utilities which direct conscious imitation in each case, by the individual creature, does not secure; while conscious imitation represents a general utility more akin to that which we have found in voluntary intelligence.

If this be so, then we have to say that conscious imitation, while it prevents the incidence of natural selection, as has been seen, and so keeps alive the creatures which have no instincts for the performance of the actions required, nevertheless does not subserve the utilities which the special instincts do, nor prevent them from having the survival value of which Romanes speaks. Accordingly, on the more general definition of intelligence, which includes in it all conscious imitation, use of maternal instruction, and that sort of thing (the vehicle of 'social transmission')—no less than on the more special definition spoken of above—we still find the principle of natural selection operative in the production of instincts and reflexes.[1]

[1] This and the two preceding papers in *Science* (and in this work) are not intended as more than preliminary statements of results thrown into the form of criticisms of particular views (*i.e.,* Romanes' and Cope's). It is for this reason that further reference is not made to the literature of the subject.

CHAPTER VII

PHYSICAL HEREDITY AND SOCIAL TRANSMISSION [1]

THE main question at issue is the relation of consciousness or intelligence to heredity, another matter, that of the relation of consciousness to the brain, being so purely speculative that it is merely touched upon at the end of this discussion.

Professor Cope [2] says: 'There is no way short of supernatural revelation by which mental education can be accomplished other than by contact with the environment through sense-impressions, and by transmission of the results to subsequent generations. The injection of consciousness into the process does not alter the case, but adds a factor which necessitates the progressive character of evolution.' Both of these sentences may be accepted, except the assertion of transmission by means of Lamarckian inheritance, which the presence of consciousness seems to render unnecessary. Using the more neutral word 'conservation' instead of 'transmission,' I may refer to three points on which Professor Cope criticises my views: first, the conservation of intelligent acquisitions from generation to generation; second, 'the progressive character of evolution'; and third, 'mental education' or acquisition.

[1] From the *American Naturalist*, May, 1896, p. 422; in formal reply to Professor Cope.

[2] *American Naturalist*, April, 1896, p. 343.

§ 1. *The Transmission of Intelligent Acquisitions*

First, accepting the statement of the fact of mental acquisition or 'selection through pleasure, pain, experience, association, etc.' (on which, see third below), Professor Cope cites the second paper (*Science*, March 20), in which I hold that consciousness makes acquisitions of new movements by such selections. He then says—if so, then I admit the Lamarckian factor. But not at all; it is just the point of the article to refute Romanes by showing that adaptation by intelligent selection makes the Lamarckian factor unnecessary. And in this way, *i.e.*, this sort of adaptation on the part of a creature *keeps that creature alive* by supplementing his reflex and instinctive actions, so *prevents the operation of natural selection* in his case, and gives the species time to get congenital variations in the lines that have thus proved to be useful (see cases cited).[1] Furthermore, all the resources of 'social transmission'—the handing down of intelligent acquisitions by parental instruction, imitation, gregarious life, etc.—come in directly to take the place of the physical inheritance of such adaptations. This influence Professor Cope, it is good to see, admits; although in admitting it, he does not seem to see that he is practically throwing away the Lamarckian factor. For instead of limiting this influence to human progress, we have to extend it to all animals with gregarious and family life, to all creatures that have any ability to imitate, and finally to all animals which have consciousness sufficient to enable them to make conscious adaptations themselves; for such creatures will have children able to do the same, and it is unnecessary to say that the children must inherit

[1] Italics in the original paper.

G

what their fathers did by intelligence, when they can do the same things by their own intelligence. As a matter of fact, Professor Cope is exactly the biologist to whose Lamarckism this admission is, so far as I can see, absolutely fatal; for he more than many others holds that accommodations all through the biological scale are secured by consciousness.[1] If so, then he is just the man who is obliged to extend to the utmost the possibility of the transmission also of these accommodations by means of intelligence, which, it appears, rules out the need of their transmission by physical heredity. At any rate, he is quite incorrect in saying that 'he [I] both admits and denies Lamarckism.'

To this form of argument Professor Cope appears to present no objection except one drawn from analogy. He says: 'I do not see how promiscuous variation and natural selection alone can result in progressive psychic evolution more than in structural evolution, since the former is conditioned by the latter.' As to the word 'progressive,' that question is taken up below; but as to the analogy with structural evolution, two answers come to mind. In the first place, Professor Cope is one of the biologists who hold that all structural evolution is secured by direct conscious accommodations. He says: 'Mind determines movements, and movements have determined structure or form.' If this be true, how can psychic be conditioned by structural evolution? Would not rather the structural changes depend upon the psychic ability of the creature to effect accommodations? And then, second, at this point Professor Cope assumes the Lamarckian factor in structural evolution. Later on he makes the same assumption when he says:

[1] And in this he is no doubt right; see Chapters VII. and IX. of *Mental Development*.

'But since the biologists have generally repudiated Weismannism,' etc. If this means Darwinism, my impression is that even on the purely biological side, the tendency is the other way. Lloyd Morgan has pretty well come over; Romanes took back before he died many of his arguments in favour of the Lamarckian factor; and quite recently a paleontologist, Professor Osborn, — if he is correctly reported in *Science*, April 3, 1896, p. 530, — argues against Professor Cope on this very point with very much the same sort of argument as this which is made here.[1] Yet Professor Cope will agree with me that this sort of *argumentum ex autoritate* is not very convincing.

But Professor Cope goes on to say that I 'both admit and deny Weismannism'; on the ground that 'his [my] denial of inheritance only covers the case of psychological sports.' But the connection is not evident. If Professor Cope means denial of the inheritance of acquired characters, then it is denied equally of sports and of other creatures; but it is not denied that the native 'sportness' (!) of sports tends to be transmitted. In my view the 'massiveness of front' which social progress shows (and which Professor Cope accepts), shows that in social transmission the individual is usually swamped in the general movement,

---

[1] Since this was written Professor Osborn has read a paper which confirms the statement of the text. Professor Osborn's expression 'ontogenic variations' *i.e.*, those brought out by 'environment (which includes all the atmospheric, chemical, nutritive, motor, and psychical circumstances under which the animal is reared)' seems to make these adaptations after all *constitutional.* As Professor Osborn says, this will not do for all cases; and I think it will not do for instinct, where constitutional variations without the aid of *consciousness* would not suffice (as Romanes says) to keep the animal alive while correlated variations were being perfected. But it seems to answer perfectly where intelligent or other accommodations supplement the constitutional variations in the species. See Appendix A, I.

as the individual sport is in biological progress. As a matter of fact, however, the analogy from 'sports' which Professor Cope makes does not strictly hold. For the social sport, the genius, is *sometimes* just the controlling factor in social evolution. And this is another proof that the means of transmission of intelligent adaptations is not physical heredity alone, but that they are socially handed down. It is difficult to see what Professor Cope means by saying that I 'admit and deny Weismannism,' for I have never discussed Weismannism at all. I believe in the Neo-Darwinian position plus some way of finding why variations count in what seem to be determinate directions; and for this latter the way now suggested appears better than the Lamarckian way. With many of the biologists (*e.g.*, Professor Minot) I see no proof of Weismannism (and protest mildly against being sorted with Mr. Benjamin Kidd!); yet I have no competence for such purely biological speculations as those which deal in plasms!

§ 2. *Progressive Evolution*

Second, the question as to how evolution can be made 'progressive.' Professor Cope thinks only by the theory of 'lapsed intelligence' or 'inherited habit'; for admitting that the intelligence makes selections, then they must be inherited, in order that the progress of evolution may set the way the intelligence selects. But suppose we admit intelligent selection (even in the way Professor Cope believes), still there are two influences at work to keep the direction which the intelligence selects apart from the supposed direct inheritance. There is that of social handing down by tradition, etc., the social transmission which has been above spoken of; and besides there is the survival by natural selection of those creatures having variations which intelligence can use. This puts a premium on these variations and their intelligent use in following generations. Suppose, for instance, a set of young animals some of which have variations which intelligence can use for a particular adaptation, thus keeping these individuals alive, while the others which have not these variations die off; then the next generation will not only have the same variations which intelligence can use in the same way, but will also have the intelligence to use the variations in the same way, and the result will be *about the same as if the second generation had inherited the adaptations directly.* The direction of the intelligent selection will be preserved in future generations. I think it is a good feature of Professor Cope's theory that he emphasizes the intelligent direction of evolution, and especially that he does it by appealing to the conscious accommodations of the creatures themselves ; but just by so doing he destroys the need of the Lamarckian factor. Natural selection eliminates all the creatures which have not the intelligence and the variations which the intelligence can use ; those are kept alive which have both the intelligence and the variations. They use their intelligence just as their fathers did, and besides get new intelligent accommodations, thus aiding progress again by further intelligent selection. What more is needed for progressive evolution? [1]

§ 3. *The Selective Process in Accommodation*

Third. We come now to the third point, — the method of intelligent selection, — and on this point Professor Cope

[1] I keep to 'intelligent' accommodations here ; but the same principle applies to *all adjustments made in individual development.*

does not understand my position, I think. I differ from him both in the psychology of voluntary accommodations of movement and in the view that consciousness is a sort of force directing brain currents in one way or another (for nothing short of a force could release or direct brain currents). The principle of dynamogenesis was cited in this form, *i.e.*, 'the thought of a movement tends to discharge motor energy into the channels as near as may be to those necessary for that movement' (above p. 55–56). This principle covers two facts. First, that no movement can be voluntarily carried out which has not itself been performed before and left traces of some sort in memory. These traces must come up in mind when its performance is again intended.[1] And second (and in consequence of this), that no act, whatever, can be performed by consciousness by willing movements which have never been performed before. It follows that we cannot say that consciousness, by selecting new adjustments beforehand, can make the muscles perform them. The most that many recent psychologists are inclined to claim is that by the attention one or other of alternative movements which have been performed before (or combinations of them) may be performed again ; in other words, selection is among old alternative movements. But this is not what Professor Cope seems to mean, nor what his theory requires. His theory requires the acquisition of new movements, *new accommodations to*

---

[1] This is formulated in the principle of ' Kinæsthetic Equivalents,' defined in the writer's *Dict. of Philos. and Psychol.* as follows : ' any mental content of the kinæsthetic order [*i.e.*, representing earlier experiences of movement] which is adequate to secure the voluntary performance of a movement. . . . The term equivalent is recommended to sum up the formulation that unless a kinæsthetic content " equivalent " to a movement be reinstated in consciousness the voluntary performance of that movement is impossible.'

environment, by a conscious selection beforehand of certain movements which are *then and for the first time carried out by the muscles.*[1]

It may very justly be asked: If his view be not true, how then can new movements which are adaptive, ever be learned at all? This is one of the most important questions, in my view, both for biologists and for psychologists; and the recent work on *Mental Development* is, in its theoretical portion (Chap. VII. ff.), devoted mainly to it, *i.e.,* the problem of *ontogenic accommodation.* We cannot go into details here, but it may suffice to say that Spencer (and Bain after him) laid out what seems to be, with certain modifications urged in that work, the only theory which can stand in court. Its main thought is this, that all new movements which are adaptive or 'fit' are selected from *overproduced movements, or movement variations,* just as organisms are selected from overproduced variations by the natural selection of those which are fit. This process, thus conceived, is there called 'functional selection,' a phrase which emphasizes the fact that it is the organism which secures from all its overproduced movements those which are adaptive and beneficial. The part which the intelligence plays 'through pleasure, pain,[2] experience, association,' etc., is to concentrate the energies of movement upon the limb or system of muscles to be used and to hold the adaptive movement, 'select' it, when it has once been struck. In the higher forms of mind both the concentration and the selection are felt as acts of attention.

---

[1] 'Conscious states do have a causal relation to the other organic processes.' I do not find, however, that Professor Cope has made clear just how in his opinion the 'selection' by consciousness works.

[2] The rôle of pleasure and pain, in regulating the discharges by a 'circular reaction,' is spoken of below, Chap. VIII. § 6.

Such a view extends the application of the general principle of selection through fitness *to the activities of the organism.* After years of study and experiment with children, etc., devoted to this problem, the writer is convinced that this 'functional selection' bears much the same relation to the doctrine of the special creation of ontogenic accommodations by consciousness which Professor Cope is reviving, that the Darwinian theory of natural selection bears to the special creation theory of the phylogenetic adaptations of species. The facts which Spencer called 'heightened discharge' are capable of formulation of the principle of 'motor excess': 'the accommodation of an organism to a new stimulation is secured — not by the selection of this stimulation beforehand (nor of the necessary movements) — but by the reinstatement of it by a discharge of the energies of the organism, concentrated, as far as may be, for the excessive stimulation of the organs (muscles, etc.), most nearly fitted by former habit to get this stimulation again,'[1] in which the word 'stimulation' stands for the condition favourable to adjustment. After several trials, with grotesquely excessive movements, the child, for example, gets the accommodation aimed at, more and more perfectly, and the accompanying excessive and useless movements fall away. This is the kind of 'selecting' that consciousness does in its acquisition of new movements. And how the results of it are conserved from generation to generation, without the Lamarckian factor, has been spoken of above.

Finally, a word merely of the relation of consciousness to the energies of the brain. It is clear that this doctrine

[1] *Mental Development*, p. 179. Spencer and Bain hold that the selection is of purely chance adjustments among spontaneous movements.

of selection as applied to muscular movement does away with all necessity for holding that consciousness even directs brain energy. The need of such direction seems to me to be as artificial as Darwin's principle showed the need of special creation to be for the teleological adaptations of the different species. This necessity of supposed directive agency done away in this case as in that, the question of the relation of consciousness to the brain becomes a metaphysical one — just as that of teleology in nature became a metaphysical one — and science can get along without asking it.[1] And biological as well as psychological science should be glad that it is so.

We may add in closing that of the three headings of this note only the last (third) is based on matters of personal opinion; the other two rest on Professor Cope's own presuppositions — that of intelligent selection in his sense of the term, and that of the bearing of social heredity (which he admits) upon Lamarckism.

[1] See the remarks on this question, below, Chap. IX. § 3.

CHAPTER VIII

A Factor in Evolution: Organic Selection [1]

In several recent publications [2] some considerations are developed, from different points of view, which tend to bring out a certain influence at work in organic evolution which we may venture to call a 'factor.' The object of the present paper is to gather into one sketch an outline of the view of the process of evolution which these different publications have hinged upon.

The problems involved in a theory of organic evolution may be gathered up under three great heads: Ontogeny or the individual's development, Phylogeny or the evolution of species, and Heredity. The general consideration, the 'factor' which it is proposed to bring out, is operative in the first instance, in the field of *Ontogeny;* I shall consequently speak first of the problem of Ontogeny; then of that of Phylogeny, in so far as the topic dealt with makes it necessary; then of that of Heredity, under the same limitation; and finally, give some definitions and conclusions.

---

[1] From the *American Naturalist,* June and July, 1896, article entitled ' A New Factor in Evolution.' Slightly revised as to terminology mainly, in accordance with the recommendations of the biological authorities of the writer's *Dictionary of Philosophy (sub verbis).*

[2] Preceding papers in this work. This essay was written to gather together the various points of view of the earlier papers, hence the frequent quotations from them.

§ 1. *Ontogenic Agencies*

Ontogeny. — The series of facts which investigation in this field has to deal with are those of the individual creature's development, and two sorts of facts may be distinguished from the point of view of the *functions which an organism performs in the course of its life history.* There is, in the first place, the development of his hereditary impulse, the unfolding of its heredity in the forms and functions which characterize its kind, together with the congenital variations which characterize the particular individual — the variations peculiar and constitutional to him — and there is, in the second place, the series of functions, acts, etc., *which he learns for himself in the course of his life.* All of these latter, *the special modifications which an organism undergoes during its ontogeny,* thrown together, have been called 'acquired characters,' and we may use that expression or adopt one recently suggested by Osborn,[1] 'ontogenic variations' (except that I should prefer the form 'ontogenetic variations') if the word 'variations' seems appropriate at all.[2]

Assuming that there are such new or modified functions, in the first instance, and such 'acquired characters' aris-

---

[1] Reported in *Science*, April 3; also used by him before the New York Academy of Science, April 13. There is some confusion between the two terminations, 'genic' and 'genetic.' I think the proper distinction is that which reserves the former, 'genic,' for application in cases in which the word to which it is affixed qualifies a term used *actively*, while the other, 'genetic,' conveys similarly a *passive* signification; thus agencies, causes, influences, etc., are 'ontogenic, phylogenic, etc.,' while effects, consequences, etc., are 'ontogenetic, phylogenetic, etc.' On terminology, see, however, the short paper reprinted below as Chap. XI. § 1.

[2] As it does not. The term modification, used above, is also given this meaning by Lloyd Morgan (*Habit and Instinct*, 1897) and is now widely adopted.

ing by 'use and disuse' from these new functions, our
further question is about them. And the question is this:
How does an organism come to be modified during its life
history?

In answer to this question we find that there are three
different sorts of ontogenic agencies which should be distinguished — each of which works to produce ontogenetic
modifications or accommodations. These are: first, the
physical agencies and influences in the environment which
work upon the organism to produce modifications of its
form and functions. They include all chemical agents,
strains, contacts, hinderances to growth, temperature
changes, etc. So far as these forces work changes in
the organism, the changes may be considered largely
'fortuitous' or accidental.[1] Considering the nature of the
forces which produce them, I propose to call these modifications 'physico-genetic.' Spencer's theory of ontogenetic
development rests largely upon the occurrence of lucky
movements brought out by such accidental influences.

Second, there is a class of modifications, in addition
to those mentioned, which arise from the spontaneous
activities of the organism itself in the carrying out of its
normal life-functions. These modifications and adjustments are seen to a remarkable extent in plants, in unicellular creatures, in very young children. There seem
to be a readiness and a capacity on the part of the organism to 'rise to the occasion,' as it were, and make gain
out of the circumstances of its life. The facts have been
put in evidence (for plants) by Henslow, Pfeffer, Sachs;
(for micro-organisms) by Binet, Bunge; (in human pathol-

---

[1] That is, so far as any direct provision for them is found in the economy
of the organism's growth.

ogy) by Bernheim, Janet; (in children) among others by the present writer (in *Mental Development,* Chap. IX., with citations ; see also Orr, *Theory of Development,* Chap. IV.). These changes I propose to call 'neuro-genetic,' laying emphasis on what is called by Romanes, Ll. Morgan, and others the 'selective property' of the nervous system, and of life generally.

Third, there is the great and remarkable series of ac-commodations secured by conscious agency, which we may throw together as 'psycho-genetic.' The processes involved here are all classed broadly under the term 'intelligent,' *e.g.,* imitation, gregarious habits, parental instruction, the lessons of pleasure and pain and of. experience generally, reasoning from means to ends, etc.

We reach, therefore, the following scheme : —

Ontogenetic Modifications	*Ontogenic Agencies*
1. Physico-genetic	1. Mechanical.
2. Neuro-genetic	2. Nervous.
3. Psycho-genetic	3. Intelligent.
	Pleasure and pain.
	Imitation.
	Higher mental processes.
	(Association of Ideas, etc.)

Now it is evident that there are two very distinct ques-tions which come up as soon as we admit modifications of function and of structure in ontogenetic development ; especially if these are considered with reference to the larger problem of evolution.

First, there is the question as to how these modifications can become adaptive in the life of the individual creature ;

or, in other words : What is the method of the individual's growth and accommodation as shown in the well-known effects of 'use and disuse'? Looked at functionally, we see that the organism manages somehow to accommodate itself to conditions which are favourable, to repeat movements which are fortunate, and so to grow by the principle of use. This involves some sort of selection, from the actual modes of behaviour of certain modes — certain functions, etc. Certain other possible and actual functions and structures decay from disuse. Whatever the method of doing this may be, we may simply, at this point, claim the law of use and disuse, as applicable in ontogenetic development, and apply the phrase, 'Functional Selection,'[1] to the organism's behaviour in acquiring new modes or modifications of adaptive function with its influence on structure. The question of the method of functional selection is taken up below (§ 6, this chapter); here we simply assume what every one admits in some form, that such adjustments of function — 'accommodations' we shall henceforth call them, the processes of learning new movements, etc. — *do occur*. We then reach another question, second : What place have these accommodations in the general theory of evolution ?

§ 2. *Effects of Individual Accommodation on Development*

In the first instance, we may note the results in the creature's own private life and development.

---

[1] Now understood from the earlier pages. In the original paper, the term 'Organic Selection' was used (see note at foot of page 96) to include the individual's functional accommodations, but later on the term was restricted as in what follows.

1. *By securing adjustments, accommodations, in special circumstances, the creature is kept alive.* This is true in all the spheres of modification distinguished in the table above. The creatures which can stand the 'storm and stress' of the physical influences of the environment, and of the changes which occur in these influences *by undergoing modifications of their congenital functions or of the structures which are constitutional to them — these creatures will live; while those which cannot will not live.* In the sphere of neuro-genetic modification we find a superb series of adjustments made by lower as well as higher organisms during the course of their development (see citations in *Mental Development*, Chap. IX.; the work of Davenport, *Experimental Morphology*, is devoted largely to this subject). And in the highest sphere, that of intelligence (including the phenomena of consciousness of all kinds, experience of pleasure and pain, imitation, etc.), we find individual accommodations on the extended scale which culminates in the skilful performances of human volition, invention, etc. The progress of the child in all the learning processes which lead him on to be a man illustrates this higher form of personal accommodation.

All these instances are associated in the higher organisms, and all of them unite to *keep the creature alive.* Passing on to consider an indirect effect of this, we find a very striking consequence.

2. By this means *those congenital or phylogenetic variations are kept in existence which lend themselves to intelligent, imitative, adaptive, or mechanical modification during the lifetime of the creatures which have them.* Other congenital variations are not thus kept in existence. So there arises a more or less widespread series of modifica-

tions *in each generation's development,*[1] *in which the con-genital and the acquired unite to produce a definite or determinate direction of change.* Those individuals in which this union of the two factors does not occur are — apart from other possible reasons for survival — incapable of maintaining the struggle for existence, and are eliminated.

The further applications of the principle lead us over into the field of our second question, that of phylogeny or evolution.

§ 3. *Effects of Individual Accommodation on Evolution*

Phylogeny: A. Physical Heredity. — The question of phylogenetic descent considered apart, in so far as may be, from that of heredity, is the question as to what the factors really are which show themselves in evolutionary progress from generation to generation. The most impor-

---

[1] "It is necessary to consider further how certain reactions of one single organism can be selected so as to adapt the organism better and give it a life history. Let us at the outset call this process 'Organic Selection' in contrast with the Natural Selection of whole organisms. . . . If this (natural selection) worked alone, every change in the environment would weed out all life except those organisms which by accidental variation reacted already in the way demanded by the changed conditions — in every case new organisms showing variations, not, in any case, new elements of life history in the old organisms. In order to the latter we should have to conceive . . . some modification of the old reactions in an organism through the influence of new conditions. . . . We are, accordingly, left to the view that the new stimulations brought by changes in the environment themselves modify the reactions of an organism. . . . The facts show that individual organisms do acquire new adaptations in their lifetime, and that is our first problem. If in solving it we find a principle which may also serve as a principle of race-development (evolution), then we may possibly use it against the 'all-sufficiency of natural selection' or in its support" (*Mental Development,* 1st ed., pp. 175–176) — quoted as an early statement (1895) of the essential idea involved in this chapter. Cf. also p. 158, below.

tant series of facts recéntly brought to light are those
which show what is called 'determinate evolution' from
one genetation to another. This has been insisted on by
the paleontologists. Of the two current theories of heredity, Neo-Lamarckism, — by means of its principle of the
inheritance of acquired characters, — has been better able
to account for this fact of determinate phylogenetic change.
Weismann admits the inadequacy of the principle of
natural selection, as operative on rival organisms, to
explain variations when they are wanted, or, as he puts
it, 'the right variations in the right place' (*Monist*, January, 1896).

It is argued in the preceding pages, that the determinate
modifications of function in ontogenesis, brought about by
neuro-genetic and psycho-genetic accommodation, do away
with the need of appealing to the Lamarckian factor. In
the case, *e.g.*, of instincts, 'if we do not assume consciousness, then natural selection is inadequate; if we
do assume consciousness, then the inheritance of acquired
characters is unnecessary' (from an earlier page). 'The intelligence which is appealed to, to take the place of instinct
and to give rise to it, uses just those partial variations
which tend in the direction of the instinct ; thus the intelligence *supplements* such partial coördinations, makes them
functional, and *so keeps the creature alive*. This prevents
the 'incidence of natural selection.' So the supposition
that intelligence is operative turns out to be just the supposition which makes use-inheritance unnecessary. Thus
kept alive, the species has all the time necessary to perfect the variations required by a complete instinct. And
when we bear in mind that the variation required is not
on the muscular side to any great extent, but in the cen-

H

tral brain connections, and is a slight variation for functional purposes at the best, the hypothesis of use-inheritance becomes not only unnecessary, but to my mind quite
superfluous' (above, Chap. V.). For adaptations generally, 'the most plastic individuals will be preserved to do
the advantageous things for which their variations show
them to be the most fit, and the next generation will show
an emphasis of just this direction in its variations' (from
an earlier page).

We get, therefore, the principle, that individual accommodations may keep a species afloat with certain results
in the sphere of phylogeny — the whole constituting the
principle of Organic Selection.

1. *It results that there arise by survival certain lines of
determinate* [1] *phylogenetic change in the directions of the determinate ontogenetic accommodations of the earlier generations.* The variations which have been utilized for ontogenetic accommodation in the earlier generations, being
thus kept in existence, are utilized more widely in the subsequent generations. 'Congenital variations, on the one
hand, are kept alive and made effective by their use for
adjustments in the life of the individual ; and, on the other
hand, adaptations become congenital by further progress
and refinement of variation in the same lines of function
as those which their acquisition by the individual called
into play. But there is no need in either case to assume
the Lamarckian factor' (from an earlier page). In cases of
conscious adaptation : 'We reach a point of view which
gives to organic evolution a sort of intelligent direction

---

[1] The phrase 'determinate change' here is merely descriptive, meaning
change in lines which keep to a definite direction. See the further discussion
of the term 'determinate' below, Chap. XII. § 1.

after all; for of all the variations tending in the direction
of an adaptation, but inadequate to its complete performance, *only those will be supplemented and kept alive
which the intelligence ratifies and uses.* The principle of
selective utility applies to the others or to some of them.
So natural selection kills off the others; and the *future
development at each stage of a species' evolution must be
in the directions thus ratified by intelligence.* So also with
imitation. Only those imitative actions of a creature
which are useful to him will survive in the species, for in
so far as he imitates actions which are injurious, he will
aid natural selection in killing himself off. So intelligence,
and the imitation which copies it, will set the direction of
the development of the complex instincts even on the
Darwinian theory; and in this sense we may say that
consciousness is a factor.'

2. *The mean of phylogenetic variations being thus made
more determinate, further phylogenetic variations follow
about this mean, and these variations are again utilized
in the process of ontogenetic accommodation.* So there is
continual phylogenetic progress in the directions set by
ontogenetic accommodation. 'The intelligence supplements slight co-adaptations and so gives them selective
utility; but it does not keep them from getting further
selective utility as instincts, reflexes, etc., by further variation' (from an earlier page). 'The imitative function, by
using muscular coördinations, supplements them, secures
accommodations, keeps the creature alive, prevents the incidence of natural selection, and so gives the species all the
time necessary to get the variations required for the full
instinctive performance of the function' (from an earlier
page). 'Conscious imitation, while it prevents the incidence

of natural selection, as has been seen, and so keeps alive
the creatures which have no instincts for the performance
of the actions required, nevertheless does not subserve the
utilities which the special instincts do, nor prevent them
from having the selective value of which Romanes speaks.
Accordingly, on the more general definition of intelligence,
which includes in it all conscious imitation, use of parental
instruction, and that sort of thing, — no less than on the
more special definition, — we still find the principle of
natural selection operative ' (from an earlier page).

3. *This completely disposes of the Lamarckian factor so
far as two lines of evidence for it are concerned.* First :
the evidence drawn from function, 'use and disuse,' is
discredited, since by organic selection the reappearance,
in subsequent generations, of the modifications first secured
in ontogenesis, is accounted for without the inheritance
of acquired characters. So also the evidence drawn from
paleontology, which cites progressive variations in the same
lines as resting on functional use and disuse. Second :
the evidence drawn from the appearance of 'determinate
variations'; for by our principle we have the continued
selection and preservation of variations in definite lines in
phylogeny without the inheritance of acquired characters.

4. *But this is not preformism in the old sense; since the
accommodations made in ontogenetic development, which
'set' the direction of evolution, are novelties of function in
whole or part* (although they utilize congenital variations of
structure). It is often by the exercise of *novel functions*
that the creatures are kept alive to propagate and thus to
produce further variations of structure which may in time
make the whole function, with its adequate structure, congenital. Romanes' arguments from 'partial co-adaptations'

and 'selective value,' seem to hold in the case of reflex and instinctive functions (see Chap. V., above), as against the old preformist or strictly Weismannist view; but the operation of organic selection, as now explained, renders these objections ineffective when urged in support of Lamarckism. 'We may imagine creatures, whose hands were used for holding only with the thumb and fingers on the same side of the object held, to have first discovered, under stress of circumstances and with variations which permitted the further adjustment, how to make use of the thumb for grasping opposite to the fingers, as we now do. Then let us suppose that this proved of such utility that all the young that did not do it were killed off; the next generation following would be plastic, intelligent, or imitative enough to do it also. They would use the same coördinations and prevent natural selection getting its work in upon them; and so instinctive "thumb-grasping" might be waited for indefinitely by the species and then be got as an instinct altogether apart from use-inheritance' (from an earlier page).[1]

5. It seems to the writer—though he hardly dares venture into a field belonging so strictly to the technical biologist — that *this principle might not only explain many cases of apparent widespread 'determinate variations' appearing suddenly, let us say, in fossil deposits, but the fact that variations seem often to be 'discontinuous.'* Suppose, for example, certain animals, varying in respect to a certain quality from *a* to *n* about a mean *x*. The mean *x* would be the case most likely to be preserved in fossil form, seeing that there are vastly more of them. Now suppose a

---

[1] Interesting cases of the operation of this principle have since been cited; cf. the extract from Headley, in Appendix B.

sweeping change in the environment, of such a kind that only the variations lying near the extreme n can accommodate to it and live to reproduce. The next generation would then show variations about the mean n. And the chances of fossils from this generation, and the subsequent ones, would be of creatures approximating n. Here would be a great discontinuity in the chain of descent and also a widespread prevalence of variations seeming to be in a single direction. This seems especially likely when we consider that the paleontologist does not deal with successive generations, but with widely remote periods, and the smallest lapse of time which he can take cognizance of is long enough to give the new mean of variation, n, a lot of generations in which to multiply and deposit its representative fossils. Of course this would be only the action of natural selection upon 'preformed' variations in those cases which did not involve positive changes, in structure and function, *acquired in ontogenesis;* but in so far as such ontogenetic accommodations were actually at hand, the extent of difference of the n-mean from the x-mean would be greater, and hence the resources of explanation, both of the sudden prevalence of the new type and of its discontinuity from the earlier, would be much increased. This additional resource is due to the organic selection factor.[1]

We seem to be able also to utilize all the evidence usually cited for the functional origin of specific characters and groupings of characters. So far as the Lamarckians have a strong case here, it remains as strong if organic selection be substituted for the 'inheritance of acquired

[1] A synopsis of the applications of this principle is given below, in Chap. XIII.

characters.' This is especially true where intelligent and imitative adaptations are involved, as in the case of instinct. This 'may give the reason, *e.g.*, that instincts are so often coterminous with the limits of species. Similar creatures find similar uses for their intelligence, and they also find the same imitative actions to be to their advantage. So the interaction of these conscious factors with natural selection brings it about that the structural definition which represents species, and the functional definition which represents instinct, largely keep to the same lines' (from an earlier page).

6. It seems proper, therefore, to call the principle of organic selection 'a new factor'; for it gives a method, hitherto undeveloped, of accounting for the parallelism between the progressive gains of evolution and the continued accommodations of individuals. *The ontogenetic modifications are really new, not preformed nor guaranteed in the variations with which the individual is born; and they really recur in succeeding generations, although not physically inherited.*

§ 4. *Tradition*[1]

B. Social Transmission. — There follows also another resource in the matter of evolution. In all the higher reaches of development we find certain coöperative or 'social' processes which directly supplement or add to the individual's private accommodations. In the lower forms it is called gregariousness, in man sociality, and in the lowest creatures, except plants, there are suggestions of a sort of recognition and responsive action between creatures

[1] This term has come into general use since this was written to designate what is transmitted socially, but not physically : see below, Chap. XI. § 1.

of the same species and in the same habitat. In all these cases it is evident that other living creatures constitute part of the environment of each, and many neuro-genetic and psycho-genetic accommodations have reference to or involve these other creatures. It is here that the principle of imitation gets very great significance; intelligence and volition come in also later on; and in human affairs we find social coöperation. Now it is evident that when young creatures have these imitative, intelligent, or quasi-social tendencies to any extent, they are able to pick up, *for themselves*, by imitation, instruction, experience generally, the functions which their parents and other creatures perform in their presence. This, then, is a form of ontogenetic accommodation; it aids to keep these creatures alive, and so to produce definite change in the way explained above. It is, therefore, a special, and from its wide range an extremely important, instance of the operation of the general principle of organic selection.

But it has further value: *it keeps alive a series of functions which either are not yet, or never do become, congenital at all.* It is a means of extra-organic transmission from generation to generation. It is analogous to physical heredity because (1) *it is a handing down of acquired physical functions*, while yet not by physical reproduction. And (2) *it directly influences physical heredity in the way mentioned*, i.e., it keeps certain variations alive, thus sets the direction of ontogenetic accommodation, thereby influences the direction of the available congenital variations of the next generation, and so determines phylogenetic evolution. It is accordingly called Social Heredity above, (Chap. IV.; see also the volumes cited, particularly *Social and Ethical Interpretations*, Chap. II.).

In social heredity, therefore, we have a more or less conservative, progressive atmosphere of which I think certain further remarks may be made.

1. *It secures adaptations of individuals all through the animal world.* 'Instead of limiting this influence to human life, we have to extend it to all the gregarious animals, to all the creatures that have any ability to imitate, and finally to all animals who have consciousness sufficient to enable them to make adjustments of their own ; for such creatures will have young that can do the same, and it is unnecessary to say that the children must inherit what their fathers did by intelligence, when they can do the same things by their own intelligence' (from an earlier page).

2. *It tends to set the direction of progress in evolution,* not only giving the young the adaptations which the adults already have, but also *producing adjustments which depend upon social coöperation; thus variations in the direction of sociality are selected and survive.* 'When we remember that the permanence of a habit learned by one individual is largely conditioned by the learning of the same habits by others (notably of the opposite sex) in the same environment, we see that an enormous premium must have been put on variations of a social kind — those which brought different individuals into some kind of joint action or coöperation. Wherever this appeared, not only would habits be maintained, but new variations, having all the force of double hereditary tendency, might also be expected' (from an earlier page). Why is it that a legitimate race of mulattoes does not arise and possess the Southern states ? Is it not the social repugnance to black-white marriages ? Remove or reverse *this influence of*

education, imitation, etc., and the result *on physical descent* would show in our faces, and even appear in our fossils when they are dug up long hence by the paleontologists of succeeding æons!

3. *In man it becomes the law of social evolution.* " Weismann and others have shown that the influence of animal intercourse, seen in parental instruction, imitation, gregarious coöperation, etc., is very important. Wallace dwells upon the actual facts which illustrate the 'imitative factor,' as we may call it, in the personal development of young animals. It has been argued that Spencer and others are in error in holding that social progress demands use-inheritance, since the socially acquired actions of a species, notably man, are socially handed down, giving a sort of 'social transmission' which supplements natural heredity" (from an earlier page). The social 'sport,' the genius, is often the controlling factor in social evolution. He not only sets the direction of future progress, but he may actually lift society at a bound up to a new standard of attainment.[1]

§ 5. *Concurrent Determination*

The two ways of securing development in determinate directions — the purely extra-organic way of social *transmission*, and the way by which organic selection in general (both by social and by other ontogenetic accommodations) secures the fixing of congenital variations, as described above — seem to run parallel.[2] Their conjoint

---

[1] The reader may consult the special developments in the work just cited.

[2] In *Social and Ethical Interpretations*, §§ 33 ff., an effort is made to show in detail that the ordinary antithesis between 'nature and nurture,' endowment and education, is largely artificial, since the two are in the main concurrent in direction.

influence is seen most interestingly in the complex instincts. We find in some instincts completely reflex or congenital functions which are accounted for by organic selection. In other instincts we find only partial coördinations given ready-made by heredity, and the creature actually depending upon some conscious resource (imitation, instruction, etc.) to bring the instinct into actual operation. But as we come up in the line of evolution, both processes may be present *for the same function;* the intelligence of the creature may lead him to do consciously what he also does instinctively. In these cases the additional utility gained by the double performance accounts for the duplication. It has arisen either (1) by the accumulation of congenital variations in creatures which already performed the action by individual accommodation and handed it down socially, or (2) the reverse. In the animals, the social transmission seems to be mainly useful as enabling a species to get instincts slowly by evolution in definite directions, the operation of natural selection being kept off. Social heredity is the lesser factor; it serves physical heredity. But in man, we find the reverse. Social transmission is the important factor, and the congenital equipment of instincts is actually broken up in order to allow the plasticity which the human being's social learning necessitates his having. So in all cases both factors are present, but in a sort of inverse ratio to each other. In the words of Preyer, 'the more kinds of coördinated movement an animal brings into the world, the fewer is he able to learn afterward.' The child is the animal that inherits the smallest number of congenital coördinations, but he is the one that learns the greatest number.

' It is very probable, as far as the early life of the child
may be taken as indicating the factors of evolution, that
the main function of consciousness is to enable him to
learn things which natural heredity fails to transmit; and
with the child the fact that consciousness is the essential
means of all his learning is correlated with the other fact
that the child is the very creature for which natural heredity gives few independent functions.   It is in this field
only that I venture to speak with assurance; but the same
point of view has been reached by Weismann and others
on the purely biological side.   The instinctive equipment
of the lower animals is replaced by the plasticity for
learning by consciousness.   So it seems to me that the
evidence points to some inverse ratio between the importance of consciousness as factor in evolution and the need
of the inheritance of acquired characters as such a factor '
(from an earlier page).

These two influences, therefore, furnish a double resort
against Lamarckism.   And I do not see anything in the
way of considering the fact of organic selection, from which
both these resources spring, as being a sufficient supplement to the principle of natural selection.   The relation
which it bears to natural selection, however, is a matter
of further remark below, in this chapter.

§ 6. *Functional Selection*

In the preceding discussions we have been endeavouring
to interpret facts.   By recognizing certain facts we have
reached a view which considers individual accommodation [1]

[1] Cf. the 'Subjective Selection' of Professor James Ward (with his allusion
to this paper) in his *Naturalism and Agnosticism*, Vol. I. p. 294.   As 'subjective' it is evidently limited to the ' psychogenic.'   I do not find that Professor

an important factor in evolution. Without prejudicing the statement of fact at all we may inquire into the actual working of the organism in making its functional selections or accommodations. The question is simply this: How does the organism secure, from the multitude of possible ontogenetic changes which it might and does undergo, those which are adaptive? As a matter of fact, all personal growth, all motor acquisitions made by the individual, show that it succeeds in doing this; the further question is, how? Before taking this up, it may be said with emphasis that the position taken in the foregoing pages, which simply makes the fact of ontogenetic accommodation a factor in development, is not involved in the solution of the further question as to how the accommodations are secured. But from the answer to this latter question we may get further light on the interpretation of the facts themselves. So we come to ask how 'functional selection' — the technical term for the process — actually operates in the case of a particular adjustment effected by an individual creature.

The organism has a way of doing this which seems to be peculiarly its own. The point is elaborated at such great length in one of the books referred to (*Mental Development*, Chaps. VII., XIII.) that details need not be repeated here. The summary made above (Chap. VII. § 3[1]) may also be referred to. There is a fact of physiology which, taken together with the facts of psychology, serves to indicate the method of the adjustments or accommodations of

---

Ward applied Subjective Selection explicitly to the problem of evolution in his original publication (*Encyclopædia Britannica*, 9th ed., art. 'Psychology'). — Note added 1902; cf. the additional note above, p. 48.

[1] On the 'circular reaction' involved, see Chap. IX. § 2.

the individual organism. The general fact is that the organism reacts by concentration upon the locality stimulated, for the continuation of the conditions, movements, stimulations, which are vitally beneficial, and for the cessation of the conditions, movements, stimulations, which are vitally depressing and harmful. In the case of beneficial conditions we find a general *increase of movement, an excess discharge of the energies of movement in the channels already open and habitual; and with this, on the psychological side, pleasurable consciousness and attention.* Attention to an organ is accompanied by increased vaso-motor activity, with higher muscular power, and a *general dynamogenic heightening in that organ.* The thought of a movement tends to discharge motor energy into the channels already established for the execution of that movement. By this organic concentration and excess of movement many combinations and variations are brought out, from which the advantageous and adaptive movements may be selected for their utility. These then give renewed pleasure, excite pleasurable associations, and again stimulate the attention, and *by these influences the adaptive movements thus struck are selected and held as permanent acquisitions.* This form of concentration of energy upon stimulated localities, with the resulting renewal through movement of conditions that are pleasure-giving and beneficial, and the subsequent repetitions of the movements, is called 'circular reaction.'[1] It seems to be the physiological basis of the selective property

---

[1] With the opposite (withdrawing, depressive effects) in injurious and painful conditions. This general type of reaction was described and illustrated, in a different connection, by Pflüger in 1877 in Pflüger's *Archiv f. d. ges. Physiologie,* Bd. XV. — (Note added 1902.)

which many have pointed out as characterizing and differentiating life. It characterizes the responses of the organism, however low in the scale, to stimulations — even to those of mechanical and chemical (physico-genic) nature. Pfeffer has shown such a determination of energy toward the parts stimulated even in plants. And in the higher animals it finds itself reproduced in type in the nervous reaction seen in imitation and — through processes of association, substitution, etc. — in all the higher mental acts of intelligence and volition. These have been developed phylogenetically as variations whose direction was constantly regulated by this form of adjustment in ontogenesis. If this be true, — and the biological facts seem fully to confirm it, — this is the adaptive process in all life, and this process it is with which the development of mental life has been in the main associated.

It follows, accordingly, that the three forms of ontogenetic modification distinguished above — physico-genetic, neuro-genetic, psycho-genetic — all involve the sort of response on the part of the organism seen in this circular reaction with excess discharge ; and we reach one general method of ontogenetic accommodation upon which organic selection rests. It is stated above in another connection in these words : " The accommodation of an organism to a new stimulation is secured, not by the selection of this stimulation beforehand (nor of the necessary movements), but by the reinstatement of it by a discharge of the energies of the organism, concentrated so far as may be for the excessive stimulation of the organs (muscles, etc.) most nearly fitted by former habit to get this stimulation again (in which the 'stimulation' stands for the condition favourable to adaptation). After several trials the child, for example, effects the adjustment aimed at, even more perfectly, and the accompanying excessive and useless movements fall away. This is the kind of selection that intelligence makes in the acquisition of new movements."

Accordingly, *all ontogenetic accommodations are neurogenetic.*[1] The general law of 'motor excess' is one of *overproduction ;* from movements thus overproduced, adjustments survive ; these adjustments set the direction of development, and by their influence in securing the survival of variations secure the same determination of direction in evolution also.[2]

The advantages of this view seem to be somewhat as follows : —

1. It gives a method of the individual's accommodations of function which is *one in principle with the law of overproduction and survival now so well established in the case of competing organisms.*

2. It reduces *nervous and mental evolution to strictly parallel terms.* The intelligent use of congenital variations for functional purposes in the way indicated, puts a premium on variations which can be so used, and thus marks out lines of progress *in directions of constantly improved mental endowment.* The circular reaction which is the method of intelligent accommodation is itself liable to variation in a series of complex ways which have produced the evolution of the mental functions known as memory, imagination, conception, thought, etc. We thus reach a phylogeny of mind which proceeds in the direction set by

---

[1] Barring, of course, those violent compelling physical influences under the action of which the organism is quite helpless, so far as such results can be called adaptive.

[2] Some of the bearings of this general theory are indicated in the following chapters.

the ontogeny of mind,[1] just as on the organic side the phylogeny of the organism gets its determinate direction from the organism's ontogenetic accommodations. And since it is the one principle of organic selection working by *the same functions* to set the direction of both phylogenies, the physical and the mental, the two developments are not two, but one. Evolution is, therefore, not more biological than psychological (cf. *Mental Development*, esp. pp. 383–388, and see the detailed statement of this requirement, on any theory of evolution, above, Part I.).

3. It makes use of the relation of structure to function required by the principle of 'use and disuse.'

4. The only alternative theories of the accommodations of the individual are those of 'pure chance,' on the one hand, and a 'creative act' of consciousness, on the other hand. Pure chance is refuted by all the facts which show that the organism does not wait for chance, but goes out in movement and effects new adjustments to its environment. Furthermore, individual accommodations are determinate; they proceed in definite, progressive lines. A short study of the child will disabuse any man, I think, of the 'pure chance' theory. But the other theory, which holds that consciousness makes adjustments and modifies structures directly by its *fiat*, is contradicted by the psychology of voluntary movement. Consciousness can bring about no movement without having first an adequate experience of that movement to serve on occasion as a stimulus to the innervation of the appropriate motor centres. 'This point is no longer subject to dispute;

---

[1] Professor C. S. Minot suggests that the terms 'onto-psychic' and 'phylo-psychic' would be convenient adjectives wherewith to mark this distinction.

I

for pathological cases show that unless some adequate idea of a former movement made by the same muscles, or by association some other idea which stands for it, can be brought up in mind, the intelligence is helpless. Not only can it not make new movements; it cannot even repeat old habitual movements. So we may say that intelligent adjustment does not create coördinations; it only makes functional use of coördinations which are alternatively present already in the creature's equipment. Interpreting this in terms of congenital variations, we may say that the variations which the intelligence uses are alternative possibilities of muscular movement' (from an earlier page). The only possible way that a really new movement can be made is *by making the movements already possible so excessively and with so many varieties of combination, etc., that new adjustments are liable to occur.*

5. The problem seems to duplicate in the main the conditions which led to the formulation of the principle of natural selection. The alternatives seemed to be 'pure chance' or 'special creation.' The law of 'overproduction with survival of the fittest' came as the solution. So in this case. Let us take an example. Every child has to learn how to write. If he depended upon chance movements of his hands, he would never learn how to write. But on the other hand, he cannot write simply by willing to do so; he might will forever without effecting a 'special creation' of muscular movements. What he actually does is to *use his hand in a great many possible ways as near as he can to the way required;* and from these excessively produced movements, and after excessively varied and numerous trials, he gradually selects and fixes the slight successes made in the direction of correct writing. It is a long and

most laborious accumulation of slight functional selections from overproduced movements.

6. The only resort left to the theory that consciousness is some sort of an *actus purus* is to hold that it *directs* brain discharges; but besides the objection that it is as hard to direct movement as it is to originate it (for nothing short of a physical force could release or direct brain energies), we find nothing of the kind necessary. The attention is what determines the particular movement in developed organisms, and the attention is no longer considered an *actus purus* with no brain process accompanying it. The attention is a function of memories, movements, previous organic experiences. We do not attend to a thing because we or the attention select it; but *we select it because we — consciousness and organism — find ourselves attending to it.*

§ 7. *The Relation of Organic to Natural Selection* [1]

A word on the relation of the principle of organic selection to that of natural selection. Natural selection is too often treated as a positive force. It is not a positive force; it is a negative formula. It is simply a statement of what occurs when some organisms do not have the qualifications necessary to enable them to survive in given conditions of life; while others by reason of their qualifications do survive. It does not in any way positively define these qualifications.

---

[1] The reader may well look up the interesting figure of Darwin at the conclusion of *Variation of Plants and Animals* (see the summary of his discussion with Asa Gray given by Poulton, *Charles Darwin*, p. 116) — the figure which describes natural selection as a builder using uncut stones (variations). Even though we side with Darwin, still the builder is better off if the stones are shaped and prepared for him by the screening and supplementing processes of individual accommodation.

Assuming the principle of natural selection in any case, and saying that, according to it, if an organism do not have the necessary qualifications it will be killed off, it still remains in that instance to find what the qualifications are which this organism is to have if it is to be kept alive. So we may say that *the means of survival is always an additional question* to the negative statement of the operation of natural selection.

This latter question, of course, the theory of variations aims to answer. The positive qualifications which the organism has arise as congenital variations of a kind which enable the organism to cope with the conditions of life. This is the positive side of Darwinism, as the principle of natural selection is the negative side.[1]

Now it is in relation to the theory of variations, and not in relation to that of natural selection, that organic selection has its main force. Organic selection points out *qualifications of a positive kind* which enable organisms to meet the environment and cope with it, while natural selection remains exactly what it was,—the negative law that if the organism does not succeed in living, then it dies. As formulating the place of such qualifications on the part of organisms, organic selection presents several additional features.

1. If we hold, as has been argued above, that the method of individual accommodation is always the same (that is, that it has a natural method), being always accomplished by a certain typical sort of nervous or vital process (*i.e.*, being always neuro-genetic), then we may ask whether that sort of process — and the consciousness

---

[1] See also the remarks made above, Chap. III. § 4; and the views of Headley and Conn in Appendix C.

which may go with it — may not be a variation appearing early in the phylogenetic series. It is argued elsewhere (*Mental Development,* pp. 200 ff. and 208 ff.) that this is the most probable view. Organisms that did not have some form of selective response to what was beneficial, as opposed to what was damaging, in the environment, could not have developed very far; and as soon as such a variation did appear it would have immediate preëminence. So we may say either that the selective vital property together with consciousness is a variation, or that it is a fundamental endowment of life and part of its final mystery.

2. But however that may be, whether individual accommodation by selective reaction and consciousness be considered a variation or a final aspect of life, it is in any case a *vital character of a very extraordinary kind.* It opens a great sphere for the application of the principle of natural selection upon organisms, *i.e., selection on the basis of what they do, rather than of what they are;* of the new use they make of their functions, rather than of the mere possession of certain congenital characters. A premium is set on plasticity and adaptability of function rather than on congenital fixity of structure; and this adaptability reaches its highest level in the intelligence.

3. It opens another field also for an analogous mode of selection, *i.e.,* the selection from particular overproduced and modified reactions of the organism, by which the determination of the organism's own growth and life history is secured. If the young chick imitated the old duck instead of the old hen, it would perish; it can only learn those new things which its present equipment will permit — not swimming. So the chick's own possible actions and accommodations in its lifetime have to be

selected. We have seen how it may be done by a certain competition of functions with survival of the fittest among them. But this illustrates the idea of natural selection. I do not see how Henslow, for example, can maintain — apart from 'special creation' — the so-called 'self-adaptations' which justify an attack on natural selection. Even plants must grow in determinate or 'select' directions in order to live, and their reactions are responses to stimulations from the environment.

4. So we may say, finally, that plasticity, while itself probably a congenital variation — or an original endowment, — works to secure new qualifications for the creature's survival, and its very working proceeds by securing a new application of the principle of natural selection to the possible modifications which the organism is capable of undergoing. Romanes says : ' It is impossible that heredity can have provided in advance for innovations upon or alterations in its own machinery during the lifetime of a particular individual.' To this we are obliged to reply in summing up — as I have done in another place : we reach 'just the state of things which Romanes declares impossible — heredity providing for the modification of its own machinery. Heredity not only leaves the future free for modifications, it also provides a method of life in the operation of which modifications are bound to come.'

§ 8. *Terminology*

In the matter of terminology some criticism is to be expected from the fact that several new terms have been used in this paper. Indeed, certain of these terms have already been criticised. It seems, however, that some novelty in terms is better than ambiguity in meanings. And in each case the new term is intended to mark off an exact meaning which no current term seems to express. Taking these terms in turn and attempting to define them, as they are used here, it will be seen whether in each case the special term is justified ; if not, the writer will be ready to abandon it.

Organic Selection : The process of individual accommodation considered as keeping single organisms alive, and so, by also securing the accumulation of variations, determining evolution in subsequent generations.

Organic selection is, therefore, a general principle of evolution which is a direct substitute for Lamarckian heredity in most, if not in all, instances. If it is really a new factor, then it deserves a new name, however contracted its sphere of application may finally turn out to be. The use of the word 'organic' in the phrase was suggested by the fact that the organism itself coöperates in the formation of the modifications which are effected, and also from the fact that, in the results, the organism is itself selected, since those organisms which do not secure the modifications fall by the principle of natural selection. The word 'selection' used in the phrase is appropriate for the reason that survival in the sense of the Darwinian meaning of 'selection' is here also denoted.[1]

Social Heredity : The acquisition of functions from the social environment, also considered as a method of determining evolution. It is a form of organic selection, but it deserves a special name because of its special way of operating. It influences the direction of evolution

---

[1] The term 'organic selection' was first used in the work, *Mental Development*, 1st ed., April, 1895. (See the notes on pp. 94 and 96.)

by keeping socially adaptive creatures alive, while others which do not adapt themselves in this way are cut off. It is also a continuous influence from generation to generation. Animals may be kept alive, let us say in a given environment, by social coöperation only; these transmit this social type of variation to posterity; *thus social accommodation sets the direction of further change, and physical heredity is determined in part by this factor.* Furthermore, the process is aided all the while, from generation to generation, by the continuous chain of extra-organic or purely social transmissions. Here are adequate reasons for marking off this influence with the name which allies the phenomenon to that of physical transmission or heredity proper.

The other terms I do not care so much about. 'Phys-ico-genetic,' 'neuro-genetic,' 'psycho-genetic,' and their correlatives in 'genic,' seem to me to be convenient terms to mark distinctions which would involve long sentences without them, besides being self-explanatory. The phrase 'circular reaction' has now been welcomed as appropriate by psychologists. 'Accommodation' is much needed as meaning single individual adjustment; the biological word 'adaptation' refers more, perhaps, to racial or general adjustments.

CHAPTER IX

MIND AND BODY[1]

§ 1. *Résumé on Consciousness and Evolution*

PROFESSOR COPE'S position as to the importance of consciousness in evolution seems in the main true as far as the question of fact is concerned. I agree with him that no adequate theory of the development of organic nature can be formulated without taking conscious states into account. The fact of accommodation requires on the part of the individual organism something equivalent to what we call consciousness in ourselves. But I do not think that the need of recognizing consciousness in connection with organic functions leads at all necessarily to the view that consciousness is a *causa vera* whose modes of action do not have physiological parallel processes in the brain and nerves. The alternatives are not really two only, automatism — a theory of mechanical causation of all movement, with the inference that consciousness is a by-product of no importance — and the *vera causa* view, which makes consciousness a new form of energy injected among the activities of the brain. There is another way of looking at the question, to which I return below.

With Professor Cope's view that the recognition of

[1] Discussion (revised) with Professors James, Cope, and Ladd before the American Psychological Association at Philadelphia, Dec. 28, 1895. From *The Psychological Review*, May, 1896, article 'Consciousness and Evolution.'

consciousness as a factor in evolution requires a Neo-Lamarckian theory of heredity I am not at all in accord. Instead of finding with Professor Cope that the emphasis of conscious function in evolution makes it necessary to recognize the Lamarckian factor, I think the facts point just the other way.[1] As soon as there is much development of mind, the gregarious or social life begins; and in it we have a new way of transmitting the acquisitions of one generation to another, which tends to supersede the action — if it exists — of physical heredity in such transmission. This transmission by 'Social Heredity' (as we have called the individual's acquisitions from society through imitation, instruction, etc.) is so universal a fact with higher animals that we may reasonably say at once that the arguments for Neo-Lamarckism drawn by Mr. Spencer and others from the phenomena of human progress, at least, are completely neutralized by it. And there are facts which show that the same state of things descends below man.

It is very probable, so far as the early life of the child may be taken as indicating the factors of evolution, that the main function of consciousness is to enable him to learn things which physical heredity fails to transmit; and with the child the fact that consciousness is the essential means of all his learning is correlated with the other fact that the child is the very creature for which physical heredity gives few congenital functions. It is in this field only that I venture to speak with assurance; but the recognition of this influence has been reached by Weismann, Ll. Morgan, and others on the purely biological side.

The instinctive equipment of the lower animals is

[1] See Chap. IV., above.

replaced by the plasticity necessary for learning by consciousness. So it seems to me that the evidence points to
some inverse ratio between the importance of consciousness as factor in development and the need of the inheritance of acquired characters as factor in evolution. This
presumptive argument may be supplemented, I think, with
positive refutations of the considerations which Professor
Cope, Mr. Romanes, and others present for the view that
the transmission of functions acquired through consciousness requires the Lamarckian factor.[1]

§ 2. *Pleasure, Pain, and the Circular Reaction*

There is one omission in Professor James' excellent
division of our topic into its members — an omission whose
importance may justify my bringing up a phase of the
general question to which I think too much importance
can hardly be attached. It is, in biological phrase, the
ontogenetic question, the examination of the development
of consciousness in the individual, with a view to interpreting the results for light upon the method of evolution.

Professor Cope's emphasis on consciousness rests here,
and it is well placed. In the life history of the organism we have the problem of development actually solved
before us in detail. The biologist recognizes this in his
emphasis on embryology and also in some degree in his
paleontology. But the psychologist has not realized the
weapon he has both for biological and for psychological use
in the mental development of the child. Moreover, the
biologist no less than the psychologist must needs resort
to this field of investigation if he would finally settle the

[1] See the preceding papers.

function of consciousness in evolution. The fossils tell nothing of any such factor as consciousness. Nor does the embryo. So, as difficult as the ontogenetic question is, it is one of the really hopeful fields on both sides. I may be allowed, therefore, to give a brief summary of certain results reached by the employment of this method; especially since it will set out more fully, even in its defects and inadequacies, the general bearing of this problem.

That there is some general principle running through all the conscious adaptations of movement which the individual creature makes, is indicated by the very unity of the organism itself. The principle of Habit must be recognized in some general way which will allow the organism to do new things without utterly undoing what it has already acquired. This means that old habits must be substantially preserved *in the new functions;* that all new functions must be reached by gradual modifications. And we will all go further, I think, and say that the only way that these modifications can be got at all is through some sort of interaction of the organism with its environment. Now, as soon as we ask how the stimulations of the environment can produce new adaptive movements, we have the answer of Spencer and Bain, — an answer directly confirmed, I think, without question, by the study both of the child and of the adult, — by the selection of fit movements from excessively produced movements, *i.e.*, from *movement variations.* So granting this, we now have the further question : How do these movement variations come to be produced *when and where they are needed?*[1] And

---

[1] This is just the question that Weismann seeks to answer (in respect to the supply of morphological variations which the paleontologists require), with his

with it, the question : How does the organism *keep those movements going* which are thus selected, and *suppress* those which are useless or damaging ?

Now these two questions are the ones which the biologists fail to answer. And the force of the facts leads to the hypotheses of 'conscious force' of Cope, 'self-development' of Henslow, and 'directive tendency' or 'determinate variation' of the Neo-Lamarckians — all aspects of the new vitalism which just these positions and the facts which they rest upon are now forcing to the front. Have we anything definite, drawn from the study of the individual on the psychological side, to substitute for these confessedly vague biological phrases ? Spencer gave an answer in a general way long ago to the second of these questions, by saying that in consciousness the function of pleasure and pain is just to keep some actions or movements going and to suppress others. The evidence of this seems to me to be coextensive, actually, with the range of conscious experience, however we may be disposed to define the physiological processes which are involved in pleasure and pain. Actions which secure pleasurable experiences to the organism are determined by the pleasure to

doctrine of 'Germinal Selection' (*Monist,* January, 1896). Why are not such applications of the principle of natural selection to variations *in the parts and functions of the single organism* just as reasonable and legitimate as is the application of it to variations in separate organisms? As against 'germinal selection,' however, I may say, that in the cases in which individual accommodation sets the direction of survival of congenital variations (as supposed in earlier pages) the hypothesis of germinal selection is in so far unnecessary. Our view finds the operation of selection *on functions in ontogeny* the means of accounting for 'variations' seeming to occur 'when and where they are wanted,' while Weismann supposes competing germinal units. Cf. the comparison of the two hypotheses, both considered as supplementary to natural selection, made by Conn, *Method of Evolution,* pp. 332–333.

be repeated, and so to secure the continuance of the pleasurable conditions; and actions which get the organism into pain are by the very fact of pain inhibited and suppressed.

But as soon as we inquire more closely into the actual working of pleasure and pain reactions, we find an answer suggested to the first question also, *i.e.*, the question as to how the organism comes to make the kind and sort of movements which the environment calls for — approximately those movement variations which are required. The pleasure or pain produced by a stimulus — and by a movement also, for the utility of movement is always that it secures stimulation of this sort or that — does not lead to diffused, neutral, and characterless movements, as Spencer and Bain suppose; this is disputed no less by the infant's movements than by the actions of unicellular organisms. There are characteristic differences in vital movements wherever we find them. Even if Mr. Spencer's undifferentiated protoplasmic movements had existed, natural selection would very soon have put an end to it. There is a characteristic antithesis between movements always. Healthy, overflowing, favourable, outreaching, expansive, vital effects are associated with pleasure; and the contrary — withdrawing, depressive, contractive, decreasing vital effects are associated with pain. This is exactly the state of things which a theory of the selection of movements from overproduced movements requires, *i.e.*, that increased vitality, represented by pleasure, should give excessive movements, from which new adjustments are selected; and that decreased vitality, represented by pain, should do the reverse — draw off energy and suppress movement.

If, therefore, we say that here is a type of reaction which all vitality shows, we may give it a general descrip-

tive name, *i.e.*, the 'circular reaction,' in that its significance for evolution is that it is not a random response
in movement to all stimulations alike, but that it distinguishes in its very form and amount between stimulations which are vitally good and those which are vitally
bad, tending to retain the good stimulations and to draw
away from and so to suppress the bad. The term 'circular' is used to emphasize the way such a reaction tends to
keep itself going, over and over, by reproducing the conditions of its own stimulation. It represents habit, since
it tends to keep up old movements; but it secures new
accommodations, since it provides for the overproduction of
movement variations for the operation of selection. This
kind of selection, since it requires the direct coöperation
of the organism itself, is known as 'Functional Selection.'
It might be called 'motor' or even 'psychic' selection,
since the part of consciousness, in the form of pleasure
and pain, and — later on — experience generally, intelligence, etc., is so prominent.[1]

This is a psychological attempt to discover the method of
the individual's accommodations; it has detailed applications
in the field of the higher mental process, where imitation,
volition, etc., afford direct exemplifications of the circular
type of reaction. But if the truth of it be allowed
by the biologist for the individual's development, the
suggestion would arise from the doctrine of recapitulation
that this type of function should run through all life.
This would mean that something analogous to consciousness (as pleasure and pain, etc.) is coextensive with life,
that the vital process itself shows a fundamental differ-

---

[1] See Chap. VII. on 'The Theory of Development,' in the work, *Mental
Development in the Child and the Race* (2d ed., 1895).

ence in movements — analogous to the difference between pleasure-incited and pain-incited movements, — and that natural selection has operated upon variations in it. The biologist may say that this is too special — this difference of reaction — to be fundamental; so it may be. But then so is life special, very special![1]

Whatever we may say to such particular conclusions, they illustrate one of the topics which should be discussed by any one, biologist or psychologist, who wants to understand the factors in evolution. There are some factors revealed in ontogenesis which do not appear in the current theories of evolution. Indeed, so far beside the mark are the biologists who are discussing transmission to-day that they generally omit — except when they hit at each other — the two factors which the psychologist has to recognize: Social Transmission, for the handing down of socially acquired characters, and Functional Selection, for the accommodations of the individual organism, with[2] whatever effects they may have on subsequent evolution.

Indeed, I do not see how either theory of heredity can get along without this appeal to ontogenesis. For if we agree in denying the inheritance of acquired characters, thus throwing the emphasis on variations, still it is only by the interpretation of ontogenetic processes and characters that any general theory of variations can be reached. Either experience causes the variations, as one theory of heredity holds; or it exemplifies them, as the other theory holds; in either case, it is the only sphere

[1] See remarks made on this and other 'comparative conceptions' above, Chap. II.

[2] Yet, of course, this statement is truer of the Darwinians than of the Lamarckians.

of fact to which appeal can be made if we would understand them. So why do biologists speculate so much as to the mode of transmission of variations, when the question of the mode of use and development of them is so generally neglected?

§ 3. *Psychophysical Dualism*

The only additional point which I may claim a little time to speak of is that to which Professor James referred in describing the current doctrines of the relation of mind and body. He described the view that consciousness does not in any way interfere with the activities of the brain, as the 'automaton theory,' and spoke as if in his mind a real automatism — the view which considers the brain processes as the sufficient statement of the grounds of all voluntary movement — is the outcome of any denial of causal energy to consciousness; in other words, that there is no alternative to what is called the epi-phenomenon theory of consciousness except a theory holding that the law of conservation of physical energy is violated in voluntary movement.

Now this reduction of the possible views to two is, in my view, unnecessary and indeed impossible. In speaking of the antecedents of a voluntary movement we have to consider the entire group of phenomenal events which are always present when voluntary movement takes place; and among the phenomena really present there is the conscious state called volition. To say that the same movement could take place without this state of consciousness is to say that a lesser group of phenomenal antecedents occurs in some cases and a

K

larger group in other cases of the same event. Why not go to the other extreme and say that the brain is not necessary to voluntary movement, since volition could bring about the movement without using the nervous processes to do it with? In his posthumous book, *Mind, Motion, and Monism*, the late Mr. Romanes brings out this inadequacy of the automaton view, using the figure of an electro-magnet, which attracts iron filings only when it is magnetized by the current of electricity. If I may be allowed to develop such a figure, I should say that whatever the electricity be, the magnet is a magnet only when it attracts iron filings; to say that it might do as much without the electricity would be to deny that it is a magnet; and the proof is found in the fact simply that it does not attract iron filings when the current is not there. So the brain is not a brain when consciousness is not there; it could not produce voluntary movement, simply because, as a matter of fact, it does not. So consciousness does not, on the other hand, produce movement without a brain. The whole difficulty seems to lie, I think, in an illegitimate use of the word ' causation.' Professor Ladd seems to me to be correct in holding that such a conception as physical causation cannot be applied beyond the sphere of things in which it has become the explaining principle, *i.e.*, in the objective, external world of things. The moment we ask questions concerning a group of phenomena which include more than these things, that moment we are liable to some new statement of the law of change in the group as a whole. Such a statement is the *third alternative* in this case; and it is the problem of the metaphysics of experience to find the broader category, the final explaining principle of experience as a whole, both objective and subjective. This I do not care to discuss, but I am far from thinking that the automaton or epi-phenomenon theorist can argue his case with much force in this higher court of appeal.

The other extreme is represented by those writers who think that the revision of the law of causation can be made in the sphere of objective phenomenal action represented by the brain; and so claim that there is a violation of the principle of conservation of energy in a voluntary movement, an actual efficiency of some kind in consciousness itself for producing physical effects. This, I think, is as illegitimate as the other view. It seems to deny the results of all objective empirical science and so to sweep away the statements of law (on one side) on which the higher interpretation of the group of phenomena as a whole must be based. And it does it in favour of an equally empirical statement of law on the other side. I do not see how any result for the more complex system of events can be reached if we deny the only principles which we have in the partial groups. To do so is to attempt to interpret the objective in terms of the subjective factor in the entire group; and we reach by so doing a result which is just as partial as that which the epi-phenomenon theory reaches in the mechanical explanation. Lotze made the same mistake long ago, but his hesitations on the subject showed that he appreciated the difficulty. I agree with these writers in the claim that the mechanical view of causation cannot be used as an adequate explaining principle of the whole personality of man; but for reasons of much the same kind it seems equally true that as long as we are talking of events of the external kind, *i.e.*, of brain processes, we cannot deny what we know of these events as such.

The general state of the problem may be shown by the accompanying diagram, which will at any rate serve the modest purpose of indicating the alternatives. The upper line (*M*) of the two parallels may represent the statements on the psychological side which, on the theory of parallelism, mental science has a right to make; the lower of the parallels (*B*) then represents the corresponding series of statements made by physics and natural science, including the chemistry and physiology of the brain. Where these lines stop an upright line may be drawn

$$M \longrightarrow \qquad \qquad \longrightarrow W$$
$$B \longrightarrow$$

to indicate the setting of the problem of interpretation in which both the other series of statements claim to be true, and the further line to the right (*W*) then gives the phenomena and statements of them which we have to deal with when we come to consider man as a whole.

Now my point is that we can neither deny either of the parallel lines in dealing with the phenomena of the single line to the right, nor can we take either of them as a sufficient statement of the further problem which the line to the right proposes. To take the line representing the mechanical principles of nature and extend it alone beyond the upright is to throw out of nature the whole series of phenomena which belong in the upper parallel line and which are not capable of statement in mechanical terms. But to extend the upper line alone beyond the upright would be to claim that mechanical principles break down in their own sphere.

As to the interpretation of the single line to the right, it may always remain the problem that it now is. The best we can do is to get points of view regarding it ; and the main progress of philosophy seems to me to be in getting an adequate sense of the conditions of the problem itself. From the more humble side of psychology, I think the growth of consciousness itself may teach us how the problem comes to be set in the form of seemingly irreconcilable antinomies. The person grows both in body and in mind, and this growth always has two sides, — the side facing toward the direction from which, the 'retrospective reference,' and the side facing the direction toward which, the 'prospective reference' of growth and the consciousness of growth. The positive sciences have by their very nature to face backwards, to look retrospectively, to be 'descriptive,' as the term is used by Professor Royce — these give the lower of our parallel lines. The moral sciences, so called, on the other hand, deal with judgments, appreciations, organizations, expectations, and so represent the other, the 'prospective' mental attitude and its corresponding aspects of reality. This gives character largely to the upper one of our parallel lines. But to get a construction of the further line, the one to the right, is to hold together both these points of view — to stand at both ends of the line — at a point where description takes the place of prophecy and where reality has nothing further to add to thought. I believe for myself that the best evidence looking to the attainment of this double point of view is found just in the fact that we are able to compass both of these functions in a measure at once; and that in our own self-consciousness we have an inkling of what that ultimate point of view is like.[1]   I do not mean to bring up points in philosophy; but it is to me the very essence of such a contention in philosophy that it is a comprehension of both aspects of phenomenal reality and not the violation or denial of either of them.

[1] This general antithesis is carried out, and various inferences are made from it, in Chaps. XVIII. and XIX. below.

CHAPTER X

DETERMINATE EVOLUTION BY NATURAL AND ORGANIC SELECTION [1]

§ 1. *Criticisms of Neo-Darwinism and Neo-Lamarckism*

ADMITTING the possible truth of either of the current doctrines of heredity, called Neo-Darwinism and Neo-Lamarckism respectively, yet there are certain defects inherent in both of them. Natural selection, considered merely as a principle of survival, is admitted by all. It fails, however, (1) to account for the lines of progress shown in evolution where the variations supposed to have been selected were not of importance enough at first to keep alive the creatures having them (*i.e.*, were not of real utility). The examination of series of fossil remains, by the paleontologists, shows structures arising with very small and insignificant beginnings.[2] Further, (2) in cases where correlations of structures and functions are in question, as in the case of complex animal instincts, it is difficult to see what utility could be attached to the partial correlations which would necessarily precede the full rise of the instinct; and yet it is impossible to believe that these correlations could have arisen by the law of variation all at once as complete functions.[3] These two

---

[1] From *The Psychological Review*, July, 1897, pp. 393 ff.

[2] Cf. the statement of this objection by Osborn, *American Naturalist*, March, 1891.

[3] Cf. Romanes, *Darwin and after Darwin*, Vol. II., Chap. III.

great objections to the 'adequacy of natural selection' are so impressive that the Neo-Darwinians have felt obliged to deal with them. The first objection may be called that from 'non-useful characters,' and the latter that from 'correlated variations.' [1]

On the other hand, the doctrine of use-inheritance or Lamarckism is open in my opinion to still graver difficulties. (1) It is a pure assumption that any such inheritance takes place. The direct evidence for it is practically nothing.[2] No unequivocal case of the inheritance of the definite effects of use or disuse has yet been cited. Again (2) it proves too much, seeing that if it actually operated as a general principle it would hinder rather than advance evolution in its higher reaches. For, first, in the more variable functions of life it would produce conflicting lines of inheritance of every degree of advantage and disadvantage, and these would very largely neutralize one another, giving a sort of functional 'panmixia' of inherited habits analogous to the panmixia of variations which arises when natural selection is not operative. Again, in cases in which the functions or acquired habits are so widespread and constant as to produce similar 'set' habits in the individuals, the inheritance of these habits would produce, in a relatively constant environment, such a stereotyped series of functions, of the instinctive type, that the plasticity necessary to the acquirement of new functions to any great extent would be destroyed. This state of things is seen in the case of certain insects which live by com-

---

[1] See the discussion of them with reference to Romanes' theory of instinct, above, Chap. V.

[2] See the candid statement of Romanes, *loc. cit.*, and cf. Lloyd Morgan, *Habit and Instinct*, Chap. XIII.

plex instincts ; and however these instincts may have been acquired, they may yet be cited to show the sort of creatures which the free operation of use-inheritance would produce. Yet just this state of things would again militate against continued use-inheritance, as a general principle of evolution ; for as instinct increases, ability to learn decreases, and so each generation would have less acquisition to hand on by heredity. So use-inheritance would very soon run itself out. Further, (3) the main criticism of the principle of natural selection cited above from the paleontologists, *i.e.*, that from 'non-useful characters,' is not met by use-inheritance ; since the lines of evolution in question are frequently, as in the case of teeth and bony structures, in characters which in the early stages of their appearance are not modified, in the direction in question, by the use of them by the creatures which have them. And, finally, (4) if it can be shown that natural selection, which all admit to be in operation in any case, can be supplemented by any principle which will meet these objections better than that of use-inheritance, then such a principle may be considered in some degree a direct substitute for the Lamarckian factor.

§ 2. *Organic Selection as a Supplementary Principle*

There is another principle at work whose operation is directly supplementary to natural selection — the principle already described above under the name of *Organic Selection.*

Put very generally, this principle may be stated as follows : acquired characters, or modifications, or individual adaptations, — all that we are familiar with in the earlier papers under the term 'accommodations,' — while not directly inherited, are yet influential in determining the course of evolution indirectly. For such modifications and accommodations keep certain animals alive, in this way screen the variations which they represent from the action of natural selection, and so allow new variations in the same directions to arise in the next and following generations; while variations in other directions are not thus kept alive and so are lost. The species will therefore make progress in the same directions as those first marked out by the acquired modifications, and will gradually 'pick up,' by congenital variation, the same characters which were at first only individually acquired. The result will be the same as to these characters, as if they had been directly inherited, and the appearance of such heredity in these cases, at least, will be fully explained; while the long-continued operation of the principle will account for 'determinate or definite' lines of change.

This principle comes to mediate to a considerable degree between the two rival theories, since it goes far to meet the objections to both of them. In the first place, the two great objections as stated above to the current natural selection theory are met by it. (1) The 'determinate' direction in evolution is secured by the indirect directive influence of organic selection — at any rate, in cases in which the direction which evolution takes is the same as that which was taken by individual modifications in earlier generations. For where the variations in the early stages of the character in question were not of utility, there we may suppose the individual accommodations to have supplemented them and so kept them in existence. An instance is seen in the fact that young chicks and ducks, which have no instinct to take up water when they see it,[1] and would perish if dependent upon the congenital variations which they have, nevertheless imitate the mother fowl, and thus, by supplementing their congenital equipment, are so kept alive.   In other fowls the drinking instinct has gone on to perfection and become self-acting. Here the accommodation secured by imitation saves the species — apart from their getting water at first accidentally — and directs its future evolution.   Further, (2) in cases of 'correlated variations' — the second objection urged above to the exclusive operation of natural selection — the same influence of organic selection is seen.   For the variations which are not adequate at first, or are only partially correlated, are supplemented by the accommodations which the creature makes, and so the species has the time to perfect its inadequate congenital mechanism.   On this hypothesis it is no longer an objection to the theory of the origin of complex instincts without use-inheritance, that these complex correlations could not have come into existence all at once;   since this principle gives the species time to accumulate and perfect its organization of them.

Similarly, the objections cited above to the theory of use-inheritance cannot be brought against organic selection.   In the first place (1) the more trivial and varied experiences of individuals — such as bodily mutilations, etc. — which it is not desirable to perpetuate, whether good or bad in themselves, would not be taken up in the evolution of the race, since organic selection would set a premium only on the variations which were important enough to be of some material use or on such as were

---

[1] See Ll. Morgan, *Habit and Instinct*, pp. 44 f., and his citations from Eimer, Spalding, and Mills.

correlated with them. These being of such importance, the species would accumulate variations in this character, and the individuals would be relieved of the necessity of making the private accommodations over again in each generation. Again (2) there would be no tendency to the exclusive production of reflexes, as would be the case under use-inheritance; since in cases in which the continued accomplishment of a function by individual accommodation was of greater utility than its accomplishment by reflexes or instinct — in these cases the former way would be perpetuated by natural selection. In the case of intelligent adaptations, for example, the increase of the intelligence with the nervous plasticity which it requires is of the greatest importance; we find that creatures having intelligence continue to acquire their adaptations intelligently with the minimum of instinctive equipment.[1] There is thus a constant interplay between instinct and accommodation, as the emergencies of the environment require the survival of one type of function or the other. This is illustrated by the fact that in creatures of intelligence we find sometimes both the instinctive and also the intelligent performance of the same function; each serving a separate utility.[2]

(3) The remaining objection — and it holds equally of both the current views — is that arising from the cases of structures which begin in a very small way with no apparent utility — such as the bony protuberances in places where

---

[1] Groos, *Play of Animals*, Eng. trans., pp. 71 ff. (see also his *Play of Man*, Eng. trans., pp. 283 f., where he admits the contention that the reverse may also be the case), has pointed out the function of imitation as aiding the growth of intelligence with the breaking up of instincts under the operation of natural selection. (See the passages in Chap. II. § 4 and Chap. XIV. § 3, where this function is cited to illustrate correlated variations.)

[2] See the statement above, Chap. VI. § 1, on 'Duplicated Functions.'

horns afterwards develop, and in certain small changes in the evolution of mammalian teeth — and afterwards progress regularly from one generation to another until they become of some utility. While it is not clear that organic selection completely accounts for these cases, yet it is quite possible that it aids us in the matter; for the assumption is admissible that in their small beginnings these characters were correlated with useful functions or variations, which, by the operation of organic and natural selection in a progressive way secured the survival and accumulation of the former. Indeed, it is part of the imperfection of the paleontological record that the evidence of such correlations would not be preserved — say, for example, muscular adjustments such as those which Weismann cites as illustrating intra-selection. It is possible that the development of muscular adjustment and strength compensated for the wearing-off of the teeth both in individual development and in evolution — as is supposed elsewhere,[1] — although the fossil teeth taken alone would give no inkling of it.

The laws of organic correlation are so little known, while yet the correlation itself is so universal,[2] that no dogmatism is justified on either side; the less perhaps on the side of the paleontologists who assert that these cases cannot be explained by natural selection even when supplemented by organic selection; for when we inquire into the state of the evidence for the so-called 'determinate variations' which are supposed in these cases, we find that it is very precarious.[3]

---

[1] Chap. XIV. § 3, and Appendix A, note to the quotation from Osborn.

[2] Instances of it are cited in Chap. XIV., below.

[3] The only way to establish 'determinate variations' would be to examine all the individuals of a given generation in respect to a given quality, and compare their mean with *the mean of their parents — not with the mean of all*

We come to the view, therefore, that evolution from generation to generation has probably proceeded by the operation of natural selection upon variations with the assistance of the organic selection of coincident (*i.e.*, those which produce congenitally what coincides with the acquisitions of the individuals) or correlated variations. And we derive a view of the relation of ontogeny to phylogeny all through the animal series. All the influences which work to assist the animal to make adjustments or accommodations will unite to give directive determination to the course of evolution. These influences we may call 'orthoplastic' or directive influences. And the general fact that evolution has a directive determination through organic selection we may call 'Orthoplasy.' [1]

As to detailed evidence of the action of organic selection, this is not the place to present it. It is well-nigh coextensive, however, with that for natural selection; for the cases where natural selection operates to preserve creatures because they adapt themselves to their environment are everywhere to be seen, and in all such cases

the individuals of the earlier generation. For some influence, such as organic selection, might have preserved only a remnant of the earlier generation, and in this way the mean of the variations of the following generation may be shifted and give the appearance of being determinate, while the variations themselves remain indeterminate. And again, the paleontologists have no means of saying how old one of these fossil creatures had to be in order to develop the character in question. It may be that a certain age was necessary and that the variations which he finds lacking would have existed if their possessors had not fallen by natural selection before they were old enough to develop this character and deposit it with their bones.

[1] These terms are akin to 'orthogenic' and 'orthogenesis,' used by Eimer (*Verh. der Deutsch. Zool. Gesell.*, 1895); the latter are not adopted, however, for the exact meaning given above, since Eimer's view directly implicates use-inheritance and 'determinate variations,' which are not made use of here. Cf. Chap. XI. § 1, on 'Terminology.'

organic selection is operative. Positive evidence in the shape of cases is, however, to be found in the papers of the writer and others on the subject.[1]

§ 3. *The Directive Factor*

We have now found some reason for the reproduction of individual or ontogenetic accommodations in phylogeny. The truth of organic selection is quite distinct, of course, from the truth of any particular doctrine as to how the accommodations in the life of the individual are effected; it may be that there are as many ways of doing this as the usual language of daily life implies, *i.e.*, mechanical, nervous, intelligent, etc.

Yet when we come to weigh the conclusions to which our earlier discussions have brought us, and remember that the type of reaction, which is everywhere present in the individual's accommodation, is the 'circular reaction' working by functional selection from over-produced movements, we see where a real orthoplastic influence in biological progress lies. The individuals accommodate by such functional selection from over-produced movements; this keeps them alive while others die; the variations which are represented in them are thus kept in existence, and further variations are allowed in the same directions. This goes on until the accumulated variations become independent of the process of individual accommodation, as congenital endowments, instincts, etc. Thus are added to the acquisitions of the species functions the same as the accommodations secured by the individuals. So race-progress shows a

[1] See also Chap. XIV., below, and the citations in Appendix A and Appendix B.

series of adaptations which corresponds in a broad way to the series of individual accommodations.

It may be remarked also that when the intelligence has reached considerable development, as in the case of man, it will outrank all other means of individual accommodation. In intelligence and will (as has been elsewhere urged)[1] the circular form of reaction becomes highly developed, and the result then is that the intelligence and the social life which it makes possible so far control the acquisitions of life as greatly to limit the action of natural selection as a law of evolution. This may be merely indicated here; the additional note below will take the subject further in the treatment of what then becomes the means of transmission from generation to generation, a form of handing down which, in contrast with physical, is called in earlier pages 'Social Transmission.'

§ 4. *Intelligent Direction and Social Progress*

The view of biological evolution already brought out has led us to the opinion that the accommodations secured by the individuals of a species are a determining factor in the progress which the species makes, since, although we cannot hold that these accommodations, or the modifications which are effected by them, are directly inherited from father to son, nevertheless by the working of organic selection, with the subsequent accumulation of variations, the course of biological evolution is directed in the channels first marked out by individual adjustments. The means of accommodation were called above orthoplastic influences in view of the directive trend which they give to the progress of the species.

[1] In the volume, *Mental Development*, Chaps. X. to XIII.

It was also intimated, in the earlier section, that when the intelligence once comes to play an important part in the accommodations of the individuals, then we should expect that it would be the controlling factor in race-progress. This happens in two ways which may now allow of brief statement.

1. The intelligence represents the highest and most specialized form of accommodation. With it goes, on the active side, the great fact of volition, which seems to spring directly out of the imitative impulse of the child. It therefore becomes the goal of organic fitness to secure the best intelligence. On the organic side, intelligence is correlated with plasticity in brain structure. Thinking and willing stand for the opposite of that fixity of structure and directness of reaction which characterize the life of instinct. Progress in intelligence, therefore, represents readiness for much acquisition, together with very little congenital instinctive equipment.

It is easy to see the effects of this. The intelligence secures the widest possible range of personal adjustments, and by so doing widens the sphere of organic selection, so that *the creature which thinks has a general screen from the action of natural selection.* The struggle for existence, depending upon the physical qualities on which the animals rely, is in some degree done away with.

This means that with the growth of intelligence, creatures free themselves more and more from the direct action of natural selection. Variations of a physical kind come to have within limits an equal chance to survive. Progress then depends on the one kind of variation which represents improved intelligence — variations in brain structure with the organic correlations which favour them — more than on other kinds.

L

2. The other consideration tends in the same direction. With the intelligence comes the growth of sentiment, especially the great class of social sentiments, and their outcome the ethical and religious sentiments. The sense of personality or self, which is the kernel of intelligent growth, involves the social environment and reflects it. Now this social sense also acts, wherever it exists, as an 'orthoplastic' influence — a directive influence, through organic selection, upon the course of evolution. In the animal world it is of importance enough to have been seized upon and made instinctive. Animal association acts to screen certain groups of creatures from the direct operation of natural selection upon them as individuals.

In man the social sentiment keeps pace with his intelligence, and so enables him again to discount natural selection by coöperation with his brethren. From childhood up the individual is screened from the physical evils of the world by his fellows. So another reason appears for considering the course of evolution to be now dominated by the intelligence.

But, it may be asked, does not this render progress impossible, seeing that it is only through the operation of natural selection upon variations — even allowing for organic selection — that progress depends? This may be answered in the affirmative, so far as progress by physical heredity is concerned. Not only do we not find such progress, but the researches of Galton, Weismann, and others show that there is probably little or no progress, even in intelligence, from father to son. The great man who comes as a variation does not commonly have sons as great. Intermarriage keeps the level of intelligent endowment relatively stable, by what Galton has called 'regression.'

Yet there is progress of another kind. With intelligence comes educability. Each generation is educated in the acquisitions of earlier generations. There is in every community a greater or less mass of so-called 'tradition' which is handed down, with constant increments, from one generation to another. The young creature grows up into this tradition by the process of imitative absorption which has been called above 'social heredity.' This directly takes the place of physical heredity as a means of transmission of many of the acquisitions which are at first the result of private intelligence, and tends to free the species from its dependence upon variations — except intellectual variations, — just as the general growth of intelligence and sentiment tends to free the organism from the law of natural selection.

These general truths cannot be expanded here; they belong to the theory of social evolution. Yet they should be noted for certain reasons which are pertinent to our general topic, and which I may briefly mention.

First, it should be said that this progress in emancipation from the operation of natural selection and from dependence upon variations, is not limited to human life. It arises from the operation of the principle which has all the while given direction to organic evolution; the principle that individual accommodations set the direction of evolution, by what is called organic selection. It is only a widening of the sphere of accommodation in the way which is called intelligent, with its accompanying tendency to social life, that has produced the deflection of the stream which is so marked in human development. And as to the existence of 'tradition' with 'social transmission' among animals, recent biological research and observations are emphasizing

them both. Wallace and Hudson have pointed out the wide operation of imitation in carrying on the habits of certain species; Weismann shows the importance of tradition as against Spencer's claim that mental gains are inherited; Lloyd Morgan has observed in great detail the action of social transmission in actually keeping young fowls alive and so allowing the perpetuation of the species, and Wesley Mills has shown the imperfection of instinct in many cases, with the accompanying dependence of the creatures upon social, imitative, and intelligent action.

Second, it gives a transition from animal to human organization, and from biological to social evolution, which does not involve a break in the chain of influences already present in all the evolution of life.

CHAPTER XI

ORGANIC SELECTION : TERMINOLOGY AND CRITICISMS

§ 1. *Terminology*[1]

IN certain recent publications [2] an hypothesis has been presented which seems in some degree to mediate between the two rival theories of heredity. The point of view taken in these publications is briefly this : Assuming the operation of natural selection as currently held, and assuming also that individual organisms through adjustment acquire modifications or new characters, then the latter will exercise a directive influence on the former quite independently of any direct inheritance of acquired characters. For organisms which survive through individual modification will hand on to the next generation any 'coincident variations' (*i.e.*, congenital variations in the same direction as the individual modifications) which they

---

[1] From *Science*, April 23, 1897, and *Nature*, LV., 1897, p. 558. See also Chap. VIII. § 8.

[2] By Osborn, Ll. Morgan, and the writer; those of Osborn and Morgan are cited in Appendix A.

This statement (§ 1) has been prepared in consultation with Principal Morgan and Professor Osborn. I may express indebtedness to both of them for certain suggestions which they allow me to use and which I incorporate verbally in the text. Among them is the suggestion that 'Organic Selection' should be the title of this paper. While feeling that this coöperation gives greater weight to the communication, at the same time I am alone responsible for the publication of it. [It was this generous action on the part of both writers which led to the final use of the term 'Organic Selection.' This paper is reproduced here in full because it presents a statement reached by coöperation and subscribed to by all of the writers mentioned.]

may chance to have, and also allow further variations in the same direction. In any given series of generations, the individuals of which survive through their susceptibility to modification, there will be a gradual and cumulative development of coincident variations under the action of natural selection. The individual modification acts, in short, as a screen to perpetuate and develop congenital variations and correlated groups of these. Time is thus given to the species to develop by coincident variation characters indistinguishable from those which were due to acquired modification, and the evolution of the race will proceed in the lines marked out by private and individual accommodations. It will appear as if the modifications were directly inherited, whereas in reality they have acted as the fostering nurses of congenital variations.

It follows also that the likelihood of the occurrence of coincident variations will be greatly increased with each generation, under this 'screening' influence of modification; for the mean of the congenital variations will be shifted in the direction of the individual modification, seeing that under the operation of natural selection upon each preceding generation variations which are not coincident [or correlated] with them tend to be eliminated.[1]

Furthermore, it has recently been shown that, independently of physical heredity, there is among the animals a process by which there is secured a continuity of social environment, so that those organisms which are born into a social community, such as the animal family, accommodate themselves to the ways and habits of that community. Professor Lloyd Morgan,[2] following Weismann and Hud-

[1] This aspect of the subject has been emphasized in Chap. X., above.

[2] Introduction to *Comparative Psychology*, pp. 170, 210, and *Habit and Instinct*, pp. 183, 342.

son, has employed the term 'tradition' for the handing on of that which has been acquired by preceding generations; and I have used the phrase 'social heredity' for the accommodation of the individuals of each generation to the social environment, whereby the continuity of tradition is secured.

It appears desirable that some definite scheme of terminology should be suggested to facilitate the discussion of these problems of organic and mental evolution; and I therefore venture to submit the following : —

1. *Variation :* to be restricted to 'blastogenic' or congenital variation.

2. *Accommodation :* functional adjustment of the individual organism to its environment. This term is widely used in this sense by psychologists, and in an analogous sense by physiologists.[1]

3. *Modification* (Lloyd Morgan) : change of structure or function due to accommodation. To embrace 'ontogenic variations' (Osborn), *i.e.,* changes arising from all causes during ontogeny.

4. *Coincident Variations* (Lloyd Morgan) : variations which coincide with or are similar in direction to modifications.

5. *Organic Selection :*[2] the perpetuation and development of congenital variations in consequence of individual accommodation.

---

[1] Professor Osborn suggests that 'individual adaptation' suffices for this; but that phrase does not mark well the distinction between 'accommodation' and 'modification' [which often takes place, as in mutilation, without accommodation]. Adaptation is used currently in a loose general sense. [It is now suggested (1892) — see the writer's *Dict. of Philos. and Psychol., sub verb.* — that adaptation be limited to racial adjustments, such as reflexes, instincts, etc., in contrast with accommodation. 'Adjustment' is a convenient general term.]

[2] Used in the papers reprinted above.

6. *Orthoplasy:*[1] the directive or determining influence of organic selection in evolution.

7. *Orthoplastic Influences:*[1] all agencies of accommodation (*e.g.*, organic plasticity, imitation, intelligence, etc.), considered as directing the course of evolution through organic selection.

8. *Tradition:* the handing on of acquired habits from generation to generation (independently of physical heredity).

9. *Social Heredity:*[2] the process by which the individuals of each generation acquire the matter of tradition and grow into the habits and usages of their kind.

§ 2. *Criticisms of Organic Selection*[3]

It is fortunate that both in Professor Wesley Mills' article in *Science*, May 22, and also in a personal letter to the writer, he accepts the class of facts emphasized in the foregoing, and admits their importance (having himself before pointed out the imperfection of instinct)[4]; the point of difference between us being in their interpretation with reference to the inheritance of acquired char-

---

[1] Used in the papers reprinted above.

[2] See the last note. Professor Lloyd Morgan thinks this term unnecessary. It has the advantage, however, of falling in with the popular use of the phrases 'social heritage' and 'social inheritance.' On the other hand, 'tradition' seems quite inadequate ; as generally used it signifies that which is handed on, the material. However, we may often employ 'social transmission' (see p. 80).

[3] From *Science*, November 13, p. 724 (an informal communication).

[4] The phrase 'half-congenital,' referred to by Professors Mills and Bumpus, was used as expressive rather than as a suggestion in terminology ! Yet the equivalent 'halb' is used in the German — so halbbewusst (subconscious), etc. See Mills, *The Nature and Development of Animal Intelligence*, in which (Part IV.) he reprints his letters and those of others.

acters. The wish may be expressed — in the way of a friendly suggestion of a reciprocal kind to Professor Mills — that he take up the arguments which are advanced above to show that the Lamarckian view of heredity is not entitled to the exclusive use of the principle of use and disuse, but that evolution may profit by the accommodations of individual creatures without the inheritance of acquired characters, through what is here called organic selection, and show why they do not apply.

As to the 'newness' of the general view which is here published, that is a matter of so little importance that I refer to it only to disavow having made untoward claims. Of course, to us all 'newness' is nothing compared with 'trueness.' As to the working of what is called 'social heredity,' it does not appear that this position was called new, *i.e.*, that social influences do aid the individual in his development and enable him to keep alive. This had been taught by Wallace, and was later signalized — as a writer on the papers points out in *Nature* — by Weismann and others. What seemed to be new about social heredity, besides the name, which appeared appropriate for reasons given in the *Naturalist* articles,[1] was the use made of it to illustrate the broader principle of organic selection — which latter principle, from certain points of view, was new. A word in regard to that.

If we give up altogether the principle of modification by use and disuse, and the possibility of new adjustments in a creature's own lifetime, we must go back to the strictest preformism. But to say that such new adjustments influence phylogenetic evolution only in case they are inherited, is to go over to the theory of Lamarckism. Now

[1] Chap. VIII. § 8.   See also Chap. XIII. § 3, note.

the position is that these individual adjustments are real (*vs.* preformism), that they are not inherited (*vs.* Lamarckism), and yet that they influence evolution. These adjustments keep certain creatures alive, so put a premium on the variations which they represent, so 'determine' the direction of variation, and give the phylum time to perfect as congenital the same functions which were thus at first only private accommodations. Thus the same result may have come about in many cases as if the Lamarckian view of heredity were true. A case of special importance of this is to be seen in *intelligent accommodations*, and one of the most interesting fields of intelligent accommodations as that of *social coöperation*.[1] The general principle, therefore, *that new adjustments effected by the individual may set the direction of evolution without the inheritance of acquired characters* is what was considered new and was called organic selection (also for reasons set out of the *Naturalist* article).

Professor Cattell, writing with thorough appreciation of the principle (in *The Psychological Review*, September, 1896, p. 572), cites Darwin's doctrine of Sexual Selection as a case from the literature. I had also reflected upon this case. But Darwin, as I think — subject to correction by those more familiar with the literature — found the importance of sexual selection in the fact that it took effect directly in the pairing of mates and so influenced posterity. It does not seem that Darwin advanced the general truth that all personal adjustments which

---

[1] These are the two main cases dealt with in my articles, and to my mind the main interest attaching to the imperfection of instinct, discussed lately by various writers in these pages (*Science*), is that it shows this 'factor' at work.

were of utility — *i.e.*, which were useful enough to enable a creature to escape with his life — would bring about indirectly the sort of effect upon pairing that sexual selection brings about directly. But whether he did or not, evidently the special case of sexual selection, as thus distinguished, does not cover the entire case, and there is the same reason for giving the whole influence or 'factor' a name that Darwin had for giving a special name to the particular case of sexual selection.

In short, does not the formulation of any sort of influence which regulates the operation of natural selection really indicate a 'factor' in the whole evolution movement? Darwin formulated sexual selection as such a factor. Wallace's 'recognition-mark' theory of the origin of bright plumage in male birds is another such formulation. Organic selection formulates a general factor by which the operation of natural selection is regulated; 'newness' in any other sense I am not disposed to maintain for it.

Darwin's personal use of the principle of sexual selection, I may add, seemed to require a very high psychological development on the part of the choosing mate, the female; but the way that the principle may be generalized —although still with reference to the special case of mating — may be seen in the very interesting suggestions of Groos (*Die Spiele der Thiere*, pp. 230 ff., Eng. trans., pp. 230 ff.; made earlier by Hirn, and reprinted in his *Origins of Art*).

More than one of my critics have spoken of the relation of organic selection to natural selection. It is discussed at some length in the *Naturalist* article (see Chap. VIII. § 7 [1]). Professor Cattell says: "It is the essence of natural selec-

---

[1] See also the remarks in Chap. III. § 5.

tion that under changed environment those individuals will survive who can best adapt themselves to it." Certainly it is. But I think that the advocates of natural selection have considered as useless or uninfluential in evolution those adjustments of individuals which were not already represented in the *congenital equipment* of the individual. Certainly the tendency, at least, of the Neo-Darwinians has been to deny the influence of the principle of use and disuse on evolution — to consider it altogether a part of the machinery of Lamarckism.[1] *The influence of new adjustments, however, in determining the limits of variation in subsequent generations without appealing to the inheritance of acquired characters* — that is the combination which we have considered new, although I should not have had the courage to label it so if certain biologists familiar with the history of discussion had not so characterized it.[2]

If Romanes, for example, had thought of this answer to Lamarckism, we cannot conceive that he would still have pressed his argument for the inheritance of acquired characters drawn from the coördinated muscular movements seen in instinct ; and in this particular case — the origin of instinct — the doctrine of organic selection appears to give a new theory.[3]

So far, however, from opposing natural selection, appeal is made directly to it. The creature that can adapt itself

[1] Thus they would say : The intelligence is congenital, but the particular things learned by intelligence, not being inherited, have as such no influence on race development, except, of course, as the children also learn to do these things intelligently.

[2] See Professor Osborn's statement beginning 'What appears to be new, therefore, in Organic Selection,' cited in Appendix A.

[3] This is now stated in detail in the writer's *Story of the Mind*, Chap. III.; see also Conn, *The Method of Evolution* (1900), pp. 269 ff.

gets its value only because it is selected, as natural selection does all its selecting. Even might we say that the very ability to make personal adaptations may possibly be due to natural selection. But Professor Cattell goes too far in saying : 'If organic selection is itself a congenital variation, as Professor Baldwin indicates [as possible],[1] we are still in the *status quo* of chance variations and natural selection.' Not entirely, indeed, since the future variations are narrowed down in their range within certain limits. Say a creature is kept alive and begets young because he can adapt himself intelligently or socially, and say his mate has the same character; then the mean of variations in the next generation will tend in the same direction, as Professor Cattell himself recognizes.[2] Of course, so far as this point goes, we do 'remain ignorant as to why the individual makes suitable adaptations'; that is quite a different question, involving, it seems, for adjustments in the sphere of muscular movement, another application of natural selection, *i.e.*, to overproduced or excessive movements[3]; but we do not remain ignorant as to 'why congenital variations occur in the line of evolution,' admitting that they occur at all. And, of course, we do remain in ignorance as to why 'they [variations] are hereditary'; that again is a matter of the mechanism of heredity.

In connection with this question of 'newness' — as unprofitable as it is to dwell upon it — another remark of

[1] Cf. my *Mental Development*, pp. 172 ff., 204 ff.

[2] In the illustration he gives of organic selection, *i.e.*, of dogs becoming granivorous from feeding on grain during many generations.

[3] Criticisms of this hypothesis of Functional Selection I cannot consider here. It is now, 1901, rather widely accepted: see Lloyd Morgan, *Animal Behaviour*, and Groos, *The Play of Man*, 'Experimenting.'

Professor Cattell may be referred to. He says that it is left in doubt whether I mean to say that the principle of organic selection was stated in my book on *Mental Development*, and also that he cannot tell from his memory of the book. This is a fair question. The principle was suggested in the book, as the quotation made from pp. 175–176 of that work (above, p. 96, note) may suffice to show.

Also in speaking of the results of the individual's accommodations on evolution, it is said: 'This again is exactly the same result as if originally neutral organisms had learned each for itself. . . . The life principle has learned, but with the help of the stimulating environment and natural selection (173).' Again, in speaking directly of heredity (pp. 205 f.): 'It [Neo-Darwinism] denies that what an individual experiences in his lifetime, the gains he makes in his adaptations to his surroundings, can be transmitted to his sons. This theory, it is evident, can be held on the view of development sketched above, for granted the learning of new movements in the way which has been called organic selection . . . yet the ability to do it may be a congenital variation. . . . And all the later acquirements of individual organisms may likewise be considered only the evidence of additional variations from these earlier variations. So it is only necessary to hold to a view by which variations are cumulative [*i.e.*, the view of organic selection] to secure the same results by natural selection as would have been secured by the inheritance of acquired characters from father to son ' (see also p. 206). It may be allowed, also, in view of the charge of obscurity made by Mr. Cattell — and the appearance of which comes in part, at least, from the need of condensation — to quote from a review of *Mental Development* in the London *Speaker*. Giving an exposition of the position which the book takes (p. 207) on the subject of heredity, the reviewer says: 'If, however, creatures having the ability to make intelligent adaptations which become consolidated into habits (called 'secondary instincts') are selected for survival, it is just as if secondary instincts were acquired by actual transmission to offspring of the modifications produced in parents by the exercise of their own intelligence. Psychologists may, therefore, practically speak as if acquired mental characters were really inherited, though what is inherited may be only the ability to acquire them. Such ability, of course, natural selection would accumulate like any other variation.'

While suggested in the book, however, it is not enlarged upon, since the section on heredity was written only to show that either of the current views might be held together with the main teaching of the book.[1]

---

[1] I regret taking so much space for these personal explanations, but the editor of *Science* can spare the space, since it is he who asked the question!

CHAPTER XII

Determinate Variation and Selection [1]

A FEW remarks may be allowed on the subject discussed in the reports of the papers of Professors Osborn and Poulton on 'Organic Selection' in the issue of *Science* for October 15, 1897.[2]

§ 1. *Determinate Variation*

1. Professor Osborn's use of the phrase 'determinate variation' seems ambiguous, and the ambiguity is the more serious since it seems to me to prejudice the main contention involved in the advocacy of organic selection. The ambiguity is this: he seems to use *determinate variation* as synonymous with *determinate evolution*.[3] He says that *determinate variation* is generally accepted, and attributes that view to Professor Lloyd Morgan and to myself. But it is only *determinate evolution* that I, for my part, am able to subscribe to; and I think the same is true of Professor Morgan.

'Determinate evolution' means a consistent and uniform direction of progress in evolution, *however that progress may be secured, and whatever the causes and processes at*

---

[1] From *Science*, November 19, 1897 (with additions).

[2] Cited in Appendix A.

[3] See his discussion, *Science*, October 15, pp. 583–584, especially p. 584, column 1, and paragraph 2 of column 2.

work. Admitting 'determinate evolution,' the question as to the causes which 'determine' the evolution is nevertheless still open, and various answers have been given to it. The Neo-Lamarckians say 'use-inheritance' (as Eimer, who calls the determination secured by this means 'orthogenesis'); Weismann says 'germinal selection'; those who accept 'organic selection' say that it is a determining factor (the resulting determination of evolution being called 'orthoplasy'); others say 'determinate variation' (continued in the same direction for successive generations); Professor Osborn says, 'determinate variation' with 'organic selection.' *Determinate variation*, then, in the proper meaning of that term, is only *one way of accounting for determinate evolution*, and to the writer it is not the true way; at any rate, it is not necessarily involved in the theory of 'organic selection.'

Let us look more closely at 'determinate variation.' Supposing that by variation we mean 'congenital variation,' then we may ask : When are variations determinate? When for any reason they are distributed in a way different from that required by the law of probability or chance. The problem of determinate variations is purely one of *distribution;* and is to be investigated for each generation, quite apart from its holding for a number of successive generations (and so giving 'determinate evolution ').

Further, the possible determinateness of variation is to be distinguished carefully from the *extent* or *width* of variation. By 'extent' of variation is meant the limits of distribution of cases about their own mean ; while relative determinateness means the distribution of cases, according to some other law than that of probabilities, about a mean established for the parents in the earlier generation.

M

The question of determinate variation is : *Has any influence worked to make the mean of variation of the new generation different from that which should be expected from the characters of their parents,*[1] *whatever the extent of variation may be.*

2. The assumption of Professor Osborn (*loc. cit.*, pp. 584–585), that because certain fossils show determinate progress, — *determinate evolution,* — therefore there must have been *determinate variation,* seems to me defective logic. It is one possibility among others, certainly, but only one. And as has been said above, Chap. X. § 3, instead of being necessary as a support for organic selection, that principle comes as a new resource to diminish the probability that the variations have really been determinate in these cases. They may be cases of orthoplasy involving organic selection working as an aid to natural selection

---

[1] I expressly avoid saying what this mean is, *i.e.*, what the contribution of each parent is to the average individual of their offspring ; but the work of Galton goes far to establish it.   Much more investigation is needed on this point of making out what is indeterminate variation ; how insecure, therefore, the claim that variations are determinate !   The drift of recent statistical studies goes, however (so far as the writer can judge), directly to show that in their distribution — considered apart from their extent — variations follow the probability curve.   They are summarized by Weldon and Davenport in the Arts. on 'Variation' in the *Dict. of Philosophy and Psychology,* Vol. II. ; see also the Arts. 'Galton's Law' (of ancestral inheritance) and 'Selection' (in biology).   The following suggestions in terminology are made by the present writer in the same work (art. 'Variation,' *ad fin.*) : "In the treatment of variation, confusion arises from failure to distinguish the following forms : (*a*) 'indefinite' or 'fortuitous' or 'ataxic' (variation subject to 'chance,' or following the law of probability); (*b*) 'definite' or 'determinate' (variation following some other law than that of probability).   The latter may well be again divided into (1) 'autotaxic' (determinate variation due to intrinsic vital tendencies to development, as held by all forms of vitalism), and (2) 'taxonomic' (determinate variation caused by external causes of any sort)." — Note added 1902.

upon 'coincident' or correlative variations which are yet not determinate but fortuitous in the strict sense.

On the doctrine of natural selection, the only way to get determinate evolution is to secure the survival of a surplus or balance of variations of a particular kind in each single generation considered for itself. So the opponents of determinate evolution have brought the challenge to show that, in each particular case, such a predominance of variations in a particular direction is found. Weismann recognizes the force of this challenge, but does not see how it can be met (especially in the form urged by the paleontologists), with all his machinery, including intraselection, and so he produces the theory of 'germinal selection' to account, as he puts it, for 'variations where and when they are wanted.' But the question is one of fact: do we actually find a balance of variations in a particular direction, antecedent to the process of elimination by natural selection? Recent statistical work points directly in the opposite direction, as is said above.

Now, the point is that the view suggested under the term Orthoplasy, with organic selection, does not require determinate variations, although it results in determinate evolution. On this view the determination is secured, not by an original balance of variations in one direction, but by a shifting of the mean of variation in a certain direction through the selective results of the creature's accommodations. These not only make their own repetition secure by repeated intra-selection in each generation, as Weismann showed, but they shield and keep alive the set of variations which they in any way involve, so that in the next generation the gamut or range of variations, *while subject to the same law of indeterminate distribution* (called ' chance distribution') *as before, yet has a mean which lies further in the direction of the accommodations themselves or in lines consistent with them.* This view is, therefore, quite consonant with the negative answer which is probably to be given to the question of fact as to determinate variation.

The ancestors of the sole, for example, had one eye on each side. Let us suppose that some of them also had a certain power of adjusting the eyes by muscular strain. Now those which could do this best in the way which would bring the eyes closer together would have the better chance of life.[1] Then, in addition to the action of natural selection upon those which were born with the eyes closer together, there would be the further fact that this acquired adjustment would save the lives of the 'accommodating' soles.[2] Not only would Weismann's intraselection have play to enable each successive generation to make the same accommodation, in turn, as their fathers had done before them, but there would be a directive tendency given to the evolution of the eyes of the sole in the matter of relative position. For while, originally, the struggle had been between those which could adjust the eyes in this manner and those which could not, the survival to maturity of the accommodating ones only would bring it about that only these would be fertile, all the next generation would have the power of some accommodation, and the mean would thus be shifted in this direction. The best accommodation would always be made by those whose

[1] By reason of some advantage, such as that arising from a flat position near the bottom, with other adaptations for better concealment, as is explained below, Chap. XIV. § 3.

[2] Professor C. B. Davenport suggests in a private letter that the principle of organic selection might be described as 'the survival of the accommodating.'

variations were in the line of this adjustment of the eyes, until finally the two eyes were found on the same side. So fruitful variation and evolution is in the line set and maintained by the individual accommodations, quite in the absence of determinate variation.

§ 2. *Selections and Selection*

3. Without going into the question, it may yet be said that the position taken by Professor Poulton in the matter of the relation of natural to organic selection — that plasticity is itself due to natural selection — is, as he says, that advocated here; but I have given natural selection still further emphasis by making the 'functional selection from overproduced movements,' whereby motor accommodations are secured, itself a case of natural selection broadly understood. I have recently drawn up a table showing the various sorts of 'selection' under the distinction of 'means' and immediate 'result,' finding some fourteen sorts of selection, and venture to reprint this table here.[1]

Certain remarks may be added to which I give numbers corresponding to those topics in the table to which they respectively relate : —

4, 5, 6. By a singular coincidence M. Delage uses the phrase 'sélection organique' (*Struct. du Protoplasma,* etc., p. 732) to describe Roux' 'Struggle of the parts,' akin to functional selection. Seeing that Weismann's

---

[1] The terms in the table which relate to social evolution are fully explained in the work, *Social and Ethical Interpretations,* Index and Appendix B, where acknowledgment is made of suggestions from Professor Lloyd Morgan. Appendix B is omitted from the third edition (1902) of that work, seeing that the table is now printed here.

SORT	MEANS	RESULT
1, 2. Natural Selection.	1. Struggle for Existence. 2. Inherent Weakness.	1. 'Survival of the Fittest' Individuals. 2. Destruction of Unfit Individuals.
3. Germinal Selection	3. Struggle of Germinal Elements.	3. Survival of Fittest Germinal Elements.
4. Intra-selection.	4. Struggle of Parts.	4. Survival of Fittest Cells and Organs.
5. Functional Selection.	5. Overproduction of Movements.	5. Survival of Fittest Functions.
6. Organic Selection.	6. Accommodation, Modification, Growth Processes.	6. Survival of Accommodating and Modified Individuals.
7. Artificial Selection.	7. Choice for Planting and for Mating together.	7. Reproduction of Desirable Individuals.
8. Personal Selection.	8. Choice.	8. Employment and Survival of Socially Available Individuals.
9. Sexual Selection.	9. Conscious Selection through Display, Courting, etc.	9. Reproduction of Attractive Individuals.
10. Social Selection. [Group Selection.]	10. Social Competition of Individuals and Groups with Natural Selection.	10. Survival of Socially Fittest Individuals and Groups.
11. Social Suppression.	11. Suppression of Socially Unfittest (by Law, Custom, etc.).	11. Survival of the Socially Fit.
12. Imitative Selection, Social Generalization.	12. Imitative Propagation from Mind to Mind with Social Heredity.	12. Survival of Ideas, Customs, etc.
13. Physiological Selection.	13. Relative Infertility.	13. Survival of the Divergent.
14. Reproductive or Genetic Selection.	14. Enhanced Fertility.	14. Survival of the Most Fertile.

'Intra-selection' (4) was directly applied by him to his interpretation of Roux' 'Struggle,' Delage's phrase is not likely to have currency as a substitute for Intra-selection. As 'Functional Selection' (5) is a special means of motor accommodation, it is additional (and in a sense subordinate) to Intra-selection, since it has a *functional* reference.

7, 8, 9. A separate heading might be given to Professor Lloyd Morgan's phrase 'Conscious Selection,' but it will be seen that, as he uses it, *i.e.*, in broad antithesis to 'Natural Selection,' it really includes all those special forms of selection in which *a state of consciousness plays the selecting rôle*[1] (7, 8, 9, 11, 12). It would be ambiguous if used for cases where *natural selection operates on mental and social variations* (5, 6, 10), since it might then mean the survival of the conscious; and even when applicable, as in sexual selection (9),[2] with respect to the 'means' of the selection, it may be ambiguous with respect to the 'result' of the selection. This last ambiguity, which is brought out in the table (8, 9),[3] makes it desirable to confine the phrase 'Conscious Selection' (if used at all) to cases which result in continuance of what is desirable for consciousness or thought. 'Personal Selection' is suggested (8) for selection by human personal choice, analogous to Sexual Selection (9) and to Romanes' 'Physiological Selection' (13). Furthermore, Darwin's 'Artificial

---

[1] This, indeed, is still liable to the question as to *whose is the state of consciousness,* giving the difference (both in means and result) seen between 'Artificial' (7) and 'Sexual' (9) selection. Ward's suggestion of the phrase 'subjective selection' (*i.e.,* by consciousness) in antithesis to natural selection (*Encyclopædia Britannica,* 9th ed., Art. 'Psychology'), was earlier.

[2] Lloyd Morgan, *Habit and Instinct,* pp. 219, 271.

[3] The bird 'selects' (sexually) for the sake of the experience, and it is a secondary result that she is also thus 'selected' for mating with the male and so for continuing his attractive characters with her own characters in the offspring.

Selection' should be used, as he used it, with reference only to securing results by induced mating (his 'Methodical' as opposed to his 'Unconscious' Selection).

10, 11, 12. In all the different sorts of 'selection,' *considered as factors in progress from generation to generation, in which the laws of natural selection and physical reproduction do not operate together*, it seems extremely desirable that we qualify the word 'selection' carefully, giving to each case a name which shall apply to it alone. The cases of the preservation of individuals and groups by reason of their social endowments do illustrate natural selection with physical reproduction, and 'Social Selection' (10) is proposed for that. In the instances in which either physical heredity is not operative (12), or in which it is not the only means of transmission (11), we cannot secure clearness without new terms; for these two cases 'Social Suppression' (11) and 'Social Generalization' (12) are suggested. The phrase 'Imitative Selection' is given in the table alternately for the latter (12), seeing that the discussions of the topic usually employ the term 'Selection' and use (wrongly) the 'Natural Selection' analogy. Selection may be used also when there is no reference to race-progress (and so no danger of the misuse of the biological analogy), since it then means presumably the 'conscious choice' of psychology and of pre-Darwinian theory.

§ 3. *Isolation and Selection*[1]

Professor Hutton protests against the use of the term 'Selection' in certain cases, saying: 'Selection means the act of picking out certain objects from a number of others,

[1] From *Science*, May 6, 1898, commenting on an article by Professor W. H. Hutton, in the same journal for April 22, 1898.

and it implies that these objects are chosen for some reason or other.' In referring to the writer's views he seems to have seen the table on p. 166, in which are given several sorts of 'selection' current in the literature of evolution. Seeing that the definition given by Mr. Hutton is pre-Darwinian, and that much of the warfare which Darwin and subsequent evolutionists had to wage was precisely over this term 'selection' — leaving aside the question whether Darwin chose the term wisely or not in the first instance — it is scarcely possible now to go back to the pre-Darwinian view which Professor Hutton advocates. Indeed, he himself, in this letter, says concerning natural selection: 'The term has become so firmly established that it can well be allowed to pass if used only in Darwin's sense of advantage gained in the struggle for existence, either by the individual or by the species.'

This admitted, there is only one thing to do, that is to recognize the two general uses of the term 'Selection,' the pre-Darwinian (or conscious) Selection 'for some reason or other,' and the Darwinian (or post-Darwinian) Selection, of which *survival on grounds of utility* is the sole criterion. Now it is true enough that all sorts of confusion arise from the interchange of these two meanings of selection; and it was with a view to the correlation of the different conceptions under certain headings ('means' and 'result') that the table was drawn up. However, it was recommended that selection in the Darwinian sense be used without qualification only when the conditions of organic progress by survival are present, namely, natural selection[1] and physical heredity. These requirements the

---

[1] In saying natural selection and physical heredity, one assumes the requisite supply of variations.

different usages of the table do fulfil; so that if each has its qualifying word ('natural,' 'sexual,' 'organic,' etc.), the use of the term 'selection' is not ambiguous. Further, in selection of the pre-Darwinian sort, as defined by Professor Hutton, *whenever it is a question of organic evolution*, these two conditions are also requisite, *i.e.*, variation and heredity, as in Darwin's artificial selection. So while fully agreeing with Professor Hutton on the necessity of definition of selection, I do not see the need of taking our nomenclature back to pre-Darwinian zoölogy. Moreover, the attempt would be quite futile.

Professor Hutton goes on to say that Darwin's term 'Natural Selection' is better than 'Organic Selection.' He seems to suppose that the two are used for the same thing. As the proposer of 'Organic Selection' (and all the other users of the term, so far as I know, *e.g.*, Osborn, Poulton, Conn, Headley, etc., have given it the same meaning), the writer can say that nothing of that sort is intended. Organic selection is supplementary; it is based upon and presupposes natural selection. It recognizes the positive accommodations on the part of individual animals by which they keep themselves alive and so have an advantage over others *under the operation of natural selection*. I agree with Professor Poulton in holding[1] that, so far from coming to replace natural selection or impair our confidence in it, it does quite the reverse. But it is also claimed that it explains cases of 'determinate evolution' which are not fully explained by natural selection alone. So some such term is justified; and it is a form of 'selection' in the Darwinian sense, for it requires both

---

[1] *Science*, Oct. 15, 1897, and *Nature*, April 14, 1898, p. 556. See also Chap. XIV. § 4, and cf. the strong statement of Headley quoted in Appendix B.

natural selection and physical heredity. Moreover, it is contrasted with natural selection on a point of which Professor Hutton speaks. He says: 'Natural Selection is not truly selection, for the individuals can hardly be said to select themselves by their superior strength, cunning, or what not.' Now, 'organic selection' supposes them doing this, in an important sense. It is a sort of artificial selection *put in the hands of the animal himself —* that is, *so far as the results go.*[1]

As to 'isolation' (Professor Hutton's other topic), it is certainly important, but is Professor Hutton right in considering it a positive cause? He says: 'It is isolation which produces the new race; selection merely determines the direction the new race is to take,' and 'isolation is capable of originating new species.' But how? Suppose we isolate some senile animals, or some physiological minors, will a new race arise? The real cause in it all is reproduction, heredity, with its likenesses and its variations. Both isolation and natural selection are negative conditions: what are called in physical science 'control' conditions, of the operation of heredity. So in seeking out such principles as 'selection,' 'isolation,' etc., we are asking how heredity has been controlled, directed, diverted, in this direction or that. Isolation is as purely negative as is natural selection. Any influence which throws this and that mate together in so far isolates them from others, as has been said in a notice of Romanes' and Gulick's doctrine of isolation,[2] and inasmuch as certain of these control conditions have already been discovered and otherwise named by their discoverer as 'natural selection,' 'artificial

[1] See below, Chap. XIII. § 1.
[2] *Psychological Review*, March, 1898, p. 216 (see Appendix C).

selection,' 'sexual selection,' etc., it is both unnecessary and unwise to attempt now to call them all 'isolation.' For if everything is isolation then we have to call each case by its special name, just the same, to distinguish it from others.

There remains the question as to whether isolation, in the broad sense of the restriction of pairing to members of the same group, can result in specific differences without any help from 'selection' of any kind. If that should be proved,[1] then there would be, it would seem, justification for the term 'isolation' in evolution theory, with a meaning not already preëmpted. This Professor Hutton claims, with Romanes and Gulick.

[1] At present it is far from being proved. Cf. Professor Cockerell's review of Romanes in *Science*, April 29, 1898.

CHAPTER XIII

ORTHOPLASY[1]

THE theory of evolution which makes general use of organic selection is called Orthoplasy; it has already been sufficiently explained. It is the theory that individual modifications or accommodations supplement, protect, or screen organic characters and keep them alive until useful congenital variations arise and survive by natural selection; and that this process, combined in many cases with 'tradition,' gives direction to evolution.

§ 1. *The Factors in Orthoplasy*

The theory, it is evident, involves two factors: (1) the survival of characters which are in any way assisted by acquired modifications, etc., during periods in which, without such assistance, they would be eliminated, until (2) the appearance and selection of congenital variations which can get along without such assistance. The second factor is simply direct natural selection; and it is the first which is the characteristic feature of this theory. By the coöperation of the acquired characters, a species or race is held up against competition and destruction, while variations

---

[1] Matter revised from the writer's *Dictionary of Philosophy and Psychology*, art. 'Organic (or Indirect) Selection,' which in that work is also signed by Professor Lloyd Morgan, Professor Poulton, and Dr. G. F. Stout. This chapter may serve as a summary statement of some of the applications of which the theory is capable, and also as a partial résumé of the preceding chapters.

are being accumulated which finally render the character
or function complete enough to stand alone. Illustrations
of this 'concurrence,'—as it is called above,—between
acquired and congenital characters, have already been
given, and others are cited in quotations made from
other writers below. The definitions of different writers
show differences of emphasis (see especially those of
Osborn and Morgan given in Appendix A).

The theory is described by Headley as 'natural selection
using Lamarckian methods' (*The Problems of Evolution*,
p. 120). Groos, in expounding organic selection, says:
'When a species have, by means of accommodations,
made new life conditions for themselves, they can manage
to keep afloat until natural selection can substitute the
lifeboat heredity for the life-preserver tradition' (*The
Play of Man*, Eng. trans., p. 283).

The term 'indirect selection,'[1] which some prefer, has
reference to the way in which natural selection comes
into operation in these cases, *i.e.*, indirectly through the
saving presence of modifications, and not directly upon
variations which are useful. Poulton had used the term
indirect in its adjective form in the following: 'These
authorities justly claim that the power of the individual
to play a part in the struggle for life may constantly give
a definite trend and direction to evolution; and although
the results of purely individual response to external forces
are not hereditary, yet *indirectly* they may result in the
permanent addition of corresponding powers to the species'
(see Appendix A, III.).

The effectiveness of the method of screening and of so

[1] This term was suggested, I think, by an anonymous writer in the *Zoological Record*.

accumulating certain variations in producing well-marked types is seen in artificial selection, where certain creatures are set apart for breeding. But any influence, such as the individual's own accommodation to his environment, which is important enough to keep him and his like alive, while others go under in the struggle for existence, may be considered with reason a real cause in producing just such effects. Thus by the processes of accommodation, a weapon *analogous to artificial selection* is put into the hands of the organism itself, and the species profits by it. Headley characterizes this aspect of the case as follows : ' The creatures pilot themselves. . . . Selection ceases to be purely natural; it is in part artificial' (see below, Appendix B, I., and above, p. 171).

For example, suppose that cats catch more long-tailed rats than short-tailed rats. Natural selection would then work to reduce the length of the rats' tails. But the breeder can secure longer-tailed rats by removing the longest-tailed, in successive generations, to an environment where there are no cats. Now suppose we find that the long-tailed rats have also more intelligence than the short-tailed ones, and use it effectively in escaping from the cats, then the effects of natural selection may be reversed: the short-tailed rats will now suffer more from the cats, and *the result will be exactly the same as that produced by the breeder* — a race of longer-tailed rats. But it is due to the screening utility of the intelligent accommodations made by the rats with long tails.

§ 2. *Applications of Organic Selection*

This point of view has had especial application and development in connection with determinate evolution, with the rise of instinct, with the origin of structures lacking in apparent utility when full formed or when only partly formed, with correlated variations, coördinated muscular groups, etc., with mental and social evolution. It would seem to be a legitimate resource in the following more special cases.

(1) In cases where there is possible correlation between the organ or function whose origin is in question and a modification which is of acknowledged utility: the latter serves as screen to the undeveloped stages of the former. This is notably the case where intelligence comes into play; it screens all sorts of characters of very varied utility.

(2) In cases of 'convergence' of lines of descent: certain accommodations, common to the two lines which converge, compel the indirect selection of variations of the same sort in the two lines, so that they are brought constantly nearer to each other; so in many cases of resemblance due to similarity of function. (This is noted by Poulton, as is also resemblance due to similarity of habit and attitude, in the art 'Mimicry,' in the present writer's *Dictionary of Philosophy and Psychology*.) The unlikelihood of two or more independent origins of the same species or character by natural selection alone has often been pointed out (cf. Poulton, *Charles Darwin*, p. 56).

There are many cases in the animal world of 'analogous' organs which are yet not 'homologous,' — organs of divergent origin but of common function, and possibly of common appearance, — the rudiments of which may have owed their common and 'indirect' selection to a single more general utility.

(3). In cases of divergent or 'polytypic' evolution: a single common character being equally available as support to two different accommodations, or as coöperating factors in them, varies in both directions, and so divergent congenital characters are evolved.

Or, again, two or more different accommodations may subserve the same utility, and thus conserve different lines of variation. To escape floods, for example, some individuals of a species may learn to climb trees, while others learn to swim. This has been recognized in Gulick's 'Change of Habits' considered as a cause of segregation, and thus also of divergent evolution.[1]

(4) In cases of apparent permanent influence, upon a stock, of temporary changes of environment, as in transplantation: the direction of variation seems to be changed by the temporary environment, when there is really only the temporary 'indirect' selection of variations appropriate to the changed environment. For example, it is possible that plants undergo quick changes by indirect selection when transplanted, the effects of this selection of variations continuing a longer or shorter period *after returning to the original conditions of life*, especially when the original environment does not demand their prompt weeding out. This is one of the cases frequently cited as favouring the hypothesis of Lamarckian inheritance.

The matter may be made clear by concrete illustrations. The point is made by Lamarckians, especially by botanists, based upon alleged facts, that modifications which are produced in plants when they are transplanted into new conditions are retained in greater or less degree by the descendants when they are re-transferred back to

[1] The implications of this position, as well as of the two preceding points (1 and 2) are brought out in the following chapter.

N

the original conditions. The argument is that the effect would not continue to appear in the environment in which there is nothing external to bring it about, unless the modifications effected by the changed environment had been inherited. This is so strong a point that many who find no evidence for Lamarckianism in other cases admit that it is likely here.

Now the point is that this relative permanence of what seems to be the influence of external conditions can be explained by organic selection. For we may hold, as it is the essence of this view to hold, that the forces of the environment in such cases modify the individuals exposed to them; and these modifications shield certain lines of variations in the same direction. If the plants lived awhile in the new environment they would show this shifting of variation in that direction; each subsequent generation would thus have less change to undergo. So to the degree that the variations were distributed about a mean different from that which existed before the plants were first removed from their original habitat, to this degree the reverse process would have to take place when they are taken back to this habitat again. That is, when first taken back they would continue to show the influence of the temporary environment without actually inheriting anything directly from it. Besides the cases of fact cited by the botanists, we may refer to the instance recently brought out on the zoological side — that of sheep said to have been transferred from Ohio to Texas, where certain changes took place in their wool — spoken of in another connection (Chap. XIV. § 1).

(5) In all cases of conscious or intelligent, including social, accommodation: in these cases conscious action

directly reënforces and supplements congenital endowment at the same time that there is indirect selection of variations which intelligence finds most suited to its needs. Thus congenital tendencies and predispositions are fostered. The orthoplastic influence of family life is well illustrated by Headley (cited in Appendix B). This is seen also in the rise of many instincts for the performance of which intelligent direction has gradually become unnecessary (cf. the use of the principle in an independent way by P. Marchal in the *Rev. Scient.*, Nov. 21, 1896, p. 653, to explain the origin of the queen bee).

The principle applies also to the origin of the forms of emotional expression (*e.g.*, Darwin's classical case of the inherited fear of man by certain birds in the Oceanic Islands: see Darwin, *Descent of Man*, Chap. II.), which are thought to have been useful, and in most cases intelligent, accommodations to an environment consisting of other animals. In man also we find reactions, such as those of bashfulness, shame, etc., largely organic, whose origin it is difficult to explain in any other way, unless we admit the inheritance of acquired characters. It is also recognized that social action by animals, as for example more or less intelligent herding, was often of direct utility and caused their survival until the corresponding instincts became fixed.

It also works another way, as Professor Groos shows: an instinct is broken up and so yields to the intelligent performance of the same function, by variations toward the increased plasticity and 'educability' which intelligent action requires. In this way another objection to Darwinism is met — that which cites the difficulty of securing the modification and decay of instincts by natural selection alone.

(6) In this connection, as we have pointed out above, we find that with the rise of intelligence, broadly understood, there comes into existence an animal tradition into which the young are educated in each succeeding generation. This sets the direction of most useful attainment, and constitutes a new and higher environment. It is with reference to this, in many cases at least, that instincts both rise and decay; decay, when plasticity and continued relearning by each generation are demanded; rise, when fixed organic reactions, stereotyped by variation and selection, are of more use. So there is a constant adjustment, as the conditions of life may demand, between the intelligent actions embodied in tradition, and the instinctive actions embodied through natural selection in inherited structure; and this is the essential coöperation of the two factors, accommodation and variation, as postulated by the theory of orthoplasy. The line of acquired modification takes the lead, variations follow. This is very different from the view which relies exclusively upon the natural selection of useful variations in this or that character; for it introduces a conserving and regulating factor, — a 'blanket utility' as it is called on an earlier page, — under which various minor adaptations may be adjusted in the organism as a whole. Of course the selection of the plasticity, required by intelligence and educability, is by direct natural selection; but, inside of this, the relation of the intelligence to the specific organic characters and functions is the one of 'concurrence' which the theory of orthoplasy postulates.

(7) It is a factor of stability and persistence of type, as opposed, for example, to the fatal result of disadvantageous variations (Wallace); since the individual accom-

modations may compensate in a constantly increasing way for the loss of direct utility of the character in question. This is notably the case with intelligent accommodations. These piece out obstructed, distorted, or partial instincts or other functions, and modify the environment to secure their free play or to negative their disadvantageous results. This carries further the advantage which Weismann has claimed in his *Romanes Lecture* for Intra-selection.

(8) It is possible, indeed, that this principle may turn out to be a resource in the difficult matter of the retrogressive evolution of particular characters, and that in two ways: (1) by the fostering of variations antagonistic to the organ or function which is undergoing decay, as is pointed out under heading (5) just above (the case of intelligent action superseding instinctive); and (2) by the fostering of a function of greater utility, which gradually replaces a lesser, in connection with the same organ or structure. For example, the evolving conformation of the skull to enclose a large brain, with growing intelligence, may have required the reduction of the biting and ear-moving muscles and the essential modification of their attachment to the bones, which became possible with the reduced utility of movable ears and powerful jaws, as intelligent accommodation advanced and replaced brute force.

Furthermore, cases of *reversed selection* are made possible under the same fostering or life-sustaining accommodation, as in the case of transplantation or removal to a new environment, and then again back to the old (see the case of plants in 4 above, and of sheep below, Chapter XIV. § 1). A similar result would show itself under great natural environmental change; and reversed selection, if only partial or temporary, would leave vestigial or partially atrophied organs. Such processes would make it unnecessary to accept for such cases the very doubtful retrogressive effects attributed by Weismann to Panmixia.

(9) It secures the effectiveness of variation in certain lines, not only by keeping alive these variations from generation to generation, but also by increasing the relative number of individuals having these variations in common, until they become established in the species. It thus answers the stock objection to natural selection (cf., *e.g.*, Henslow, *Natural Science*, VI., 1895, pp. 585 f., and VIII., 1897, pp. 169 f.) which claims that the same variation would not occur at any one time in a sufficient number of individuals to establish itself, except in case of great environmental change or of migration. Organic selection shifts the mean of a character, and this changed mean is what natural selection requires (cf. Conn, *The Method of Evolution*, pp. 75 f.).

It aids, also, in the matter of discontinuous variation — first, by allying accommodation possibly with extreme variations and so making them useful; and second, by presenting the appearance of discontinuous variation, as mentioned in Chapter VIII. § 3 (5).

(10) Organic selection is a segregating or isolating factor, as is illustrated under (3) above. Animals which make common accommodations survive and mate together. In the presence of an enemy, for instance, those animals which can run fast escape together; those which can go through small holes remain likewise together; and so do those hardy enough to fight, etc. This effective production of separate groups is directly due to the different accommodations respectively made in the individuals.

§ 3. *Intra-selection and Orthoplasy*

As to the possible universal application of organic selection, which makes orthoplasy a general theory, it would seem to depend upon whether there are any cases of congenital characters maturing without some accommodation due to the action of the life conditions upon the individual's plastic material. The position is maintained above that there are probably no such characters. It follows that those variations in which the most fortunate combination of innate and acquired elements is secured survive under natural selection; and this means that organic selection is universal. In the words of Groos (*Play of Man*, Eng. trans., p. 373), 'organic selection may possibly be applied to all cases of adaptation (Anpassung).'

This point of view follows naturally from the position taken up by the school of Organicists already mentioned above,[1] who insist in various ways upon the part played by the organism itself in evolution. The writers of this school, however, either hold to Lamarckian inheritance (Eimer), to a form of self-development (Driesch, called 'auto-régulation' and 'auto-détermination' by Delage), or to Intra-selection (a term of Weismann's) considered as *repeating its results anew in each generation* (Roux, Delage). Weismann (*Romanes Lecture*) combines intra-selection, which ' effects the special adaptation of the tissues . . . *in each individual*' (' for *in each individual* the necessary adaptation will be temporarily accomplished by intra-selection ') with the hypothesis of Germinal Selec-

---

[1] Chap. III. § 1 ; cf. Herbst, *Formative Reize in Ontogenese* (1901). The least modifiable characters, such as coloration as in protective colours, mimicry, etc., would most nearly fulfil this condition.

tion. 'As the primary variations,' says Weismann, 'in the phyletic metamorphosis occurred little by little, *the secondary adaptations would probably, as a rule, be able to keep pace with them.* Time would thus be gained, till, in the course of generations, by constant selection of *those germs, the primary constituents of which are best fitted to one another* . . . a definite metamorphosis of the species, involving all the parts of the organism, may occur.' In this passage (which has been quoted by Osborn and others to show that this writer anticipated the principle of organic selection[1]) Weismann recognizes the essential coöperation of variation and modification which organic selection postulates, *but he reverses the order of these factors by making germinal variations* (in the words italicized above by the present writer) *the leading agency* in the determination of the course of evolution, while individual accommodation and modification 'probably keep pace with them' (the primary variations). The writers who originally expounded organic selection rely upon variation to 'keep pace,' under the action of natural selec-

---

[1] The two other authorities whose theories have been similarly cited are Delage and Pfeffer (see Davenport in *The Psychological Review*, IV., November, 1897, p. 676). Both these writers, however, as I read them, stop substantially with intra-selection — and 'struggle of parts'— 'repeating its results anew in each generation,' *i.e.,* with increased plasticity and continued accommodation. Delage has himself confirmed this interpretation in writing explicitly upon organic selection (see *Année Biologique*, III., 1899, p. 512). The view of Pfeffer on this point is indicated by the lines quoted by Davenport in the place cited just above to the effect that, given struggle for existence and the resulting individual modifications, 'the remaining part of the Darwinian theory, namely, the gradual production of new races and species, seems consequently unnecessary.' In other words, both these writers are 'Organicists' who do not combine that position with the natural selection of variations. Delage's work is that on *Structure du Protoplasma, &c.,* already frequently cited (see especially pp. 824–833); Pfeffer's theory is in *Verhandlungen des naturwissenschaftlichen Vereins in Hamburg*, 1893, pp. 44 ff.

tion, *with individual accommodation, which last thus takes
the lead* and marks out the course of evolution. The
hypothesis of germinal selection, which is essential to Weismann's view, is not at all involved in theirs. In the words
of Lloyd Morgan, whose account of the relation between
Weismann's views and his own (see Appendix A, II.[1])
should be turned to : ' Natural selection would work along
the lines laid down for it by adaptive modifications.[2] Modification would lead; variation follow in its wake. Weismann's germinal selection, if a *vera causa*, would be a
coöperating factor and assist in producing the requisite
variations.' Defrance says on the same point (*Année
Biol.*, III., 1899, p. 533): 'He (Weismann) has made use
of his personal hypotheses on germ-plasm which are not
universally admitted, while the conception of Lloyd Morgan
and Baldwin avoids this stumbling-block by not closing
inquiry into the processes which enter into play. It is true
that this leaves it an hypothesis; but it is nevertheless
true that it offers an intelligible solution of one of those
problems which appear on the surface to constitute the
most insoluble of enigmas.' Osborn brings into play the
further factor of ' determinate variation,' which, if true,
would be analogous in its rôle to Weismann's germinal
selection (see citation from Osborn, Appendix A, I.); he
also holds that ' there is an unknown factor in evolution
yet to be discovered.'

---

[1] Especially the ' new statement ' there given.

[2] On this positive ground, I think the term ' organic selection ' is to be pre-
ferred to ' indirect selection.' Historically, it follows Delage's use of ' Organicists ' for the school of writers who lay stress on individual accommodation;
and also his use of *sélection organique* — see above, Chapter XII. § 2 (4, 5, 6) —
for the process of accommodation shows how similar concepts may suggest an
identical term to different writers.

§ 4. *Three Types of Theory*

Any general theory of determinate and divergent evolution — in short any theory of descent, whether of a character or of an organism, of mind or of body — has a complexion derived from the composition of the factors it employs or assumes. The theory which exclusively employs natural selection is called Neo-Darwinism or Weismannism; the theory which gives use-inheritance a large place, whether laying greater or less stress upon natural selection and other factors as subordinate, are called Lamarckism or Lamarckianism. Vitalism takes on a variety of forms, which have specific names, according as their respective holders make prominent specific modes of operation in which the life forces work themselves out — as the 'self-development' theory, the auto-régulation (Delage) theory, the orthogenesis theory, the theories of bathmism, growth-energy, etc. This is certainly both legitimate and convenient — to suggest a term to designate a theory which, in its main hypothesis or in its manner of grouping the subsidiary hypotheses, presents a distinguishable and discussable whole.[1]

The theory which is expounded in these pages presents two principles which, both in the formation given to them and in the rôle assigned to them in the theory of descent, mark it as having such a distinguishable and discussable

---

[1] Professor Conn, speaking of the theories of isolation and organic selection, says: 'There can be no question that these two theories are important contributions to the problem of organic evolution. In regard to the disputed question of whether they are a part of natural selection and therefore included in Darwinism we need attempt no decision. They certainly represent aspects of the problem not recognized until recently, and may therefore be looked upon as actual contributions to our knowledge of evolution' (*The Method of Evolution*, p. 333).

character — the principles of organic selection and social transmission. These two principles, which rest upon facts for their validity, are in this theory given a place and a relation to the other factors of the theory of descent, and also to each other, not given to them in any other general theory. It is accordingly quite within the general usage of biologists to give such a theory a name. The term 'orthoplasy' has been suggested above as such a designation for the theory — a term which from its derivation (Greek ὀρθός straight, and πλάσις, a moulding — seen in the

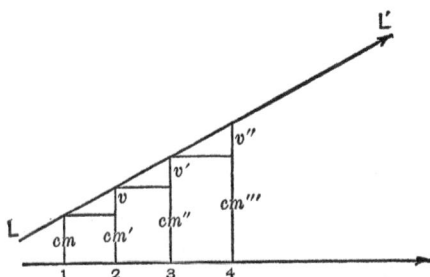

Theory of Neo-Darwinism or Weismannism. LL', line of evolution; 1, 2, etc., successive generations by physical heredity; *cm*, *cm'*, etc., congenital mean; *v*, *v'*, variations (congenital). Evolution is by natural selection of variations added to the congenital mean from generation to generation.

English word *plastic*) appropriately designates a view which mainly concerns itself with the factors at work in the determination or direction of the movement of evolution.

The relation of this theory to other current general views is indicated here and there in the preceding pages. Many of the papers here reprinted were written in the first instance to show that this theory is free from objections urged to Neo-Lamarckism and Neo-Darwinism; and it has been pointed out in what way orthoplasy finds itself 'orientated' with respect to the less general truths on which all the theories must ultimately repose. We may accordingly display, by the three cuts given herewith, the

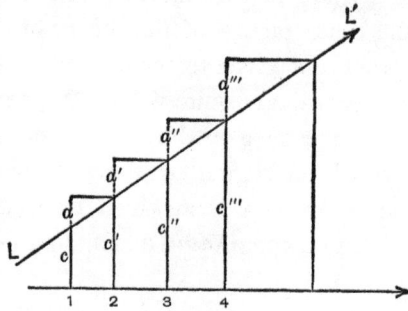

Theory of Lamarckism or Neo-Lamarckism. LL', line of evolution; 1, 2, etc., successive generations by physical heredity; *c, c'*, etc., congenital endowment; *a, a'*, etc., modifications (acquired). The modification of one generation is added to the endowment of the next by the principle of use-inheritance.

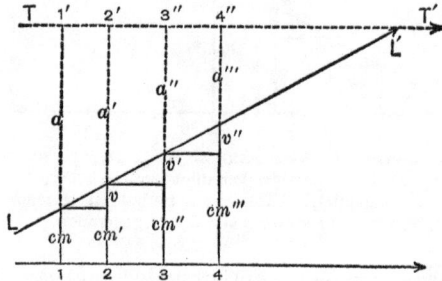

Theory of Orthoplasy. LL', line of evolution; 1, 2, etc., successive generations by physical heredity; TT', line of tradition (social transmission); *cm, cm'*, etc., congenital mean; *a, a'*, etc., accommodations (and modifications) supplementing or screening *cm*, etc.; *v, v'*, etc., (congenital) variations added to *cm*, etc., by natural selection. The species is kept alive by *a, a'*, etc., and TT', during the evolution of *cm*. The line TT', considered as 'tradition,' is of varying importance according to the character in question and to the grade of the organism in the scale of life; but if it signify any utility for which the accommodations are necessary, it is always present, and may be called the 'line of utility.'

contrasts presented by the three views when their essentials are compared with one another. The contrasts are real, as the differences in the diagrams show.

§ 5. *Concurrence and Recapitulation*

From the point of view of the theory of orthoplasy, we have a somewhat modified way of viewing the general principle of recapitulation. The statement of the intergenetic relation of evolution and development as one of 'concurrence' gives us this changed point of view; for concurrence is to be interpreted as well from the side of the leading part played by accommodation and not simply, as is the case with recapitulation, from that of stages of the ontogenetic processes of heredity. *Concurrence of the sort reached by the theory of orthoplasy* [1] *amounts to a sort of reversed recapitulation.* The individual recapitulates his genetic series, but the genetic series became what it is by reason of its continued concurrence with the processes of individual accommodation.

If we ask the philosophical question, why recapitulation is true — why development should recapitulate evolution — various partial answers may be advanced; and from the point of view of orthoplasy greater cogency and completeness seems to be given to these partial answers.

We may say, *first*, that the process which shows itself as recapitulation, is the only one by which nature could make individuals *like their parents;* the way, that is, of bringing them up through serial processes of genetic development, each stage being necessary to the next. If nature, by variation, departed too widely from this series of processes, the individuals would fail of some of the essential adaptations which just this series of genetic pro-

---

[1] In general conception, of course, a concurrence might arise from Lamarckian heredity. It is this general use of the term 'concurrence' which is contrasted with the conception of coincidence, due to coincident variation alone, which is taken up in the next chapter.

cesses represents; they would thus miss being in the same degree like their parents.   In short, *recapitulation is a sine qua non of heredity.*

This appears reasonable; and it becomes more so when we take the point of view of concurrence.   For on that view, future evolution in succeeding generations is to be in the lines marked out by accommodation in the preceding generations.   If this is also to exhibit itself in the process of development, then it is the more important that the offspring in each case should have new elements of endowment (variation), not inconsistent with the processes of development through which the parents also acquired theirs.   This would extend backward from generation to generation.   In other words, variations, to be effective in the same functional lines with accommodation, should be consistent with the processes of development already established, up to the point from which accommodations to the environment, of like nature to themselves, have been found possible.

This additional point may be put in some such way as this : not only is recapitulation a *sine qua non* of heredity, but *recapitulation plus variation is a sine qua non of heredity plus concurrence.*   Variation and modification which concur in direction are most likely from processes which are common to the two genetic series to which they respectively belong.

Referring to the diagrams given in the last paragraph (pp. 187 f.) for natural selection (Neo-Darwinism) and orthoplasy, the two points just made may be illustrated from them.   Referring to the natural selection diagram, it appears that the individual development cm''', in order to issue in an adult showing heredity from cm'', must go through processes like those of cm''; cm'' in turn through processes like those in cm'; and so on indefinitely back to the ancestors of cm. The entire series will then be reflected, apart from modifying agencies, in cm'''.

But now referring to the diagram for orthoplasy, we read the facts the other way. If the variation v is to be effective as coincident or concurrent with the modification a, then the processes cm' which lead up to v are most reasonably the same as cm which lead up to a. So the processes cm'' should be expected to repeat those of cm', those of cm''' those of cm'', etc., each meeting the requirement made upon it of affording continued support to concurrence with the continued accommodation processes a', a'', a''', etc.

It may be said that this extended way of producing an individual, by development through a series of stages, is cumbersome, and that it would be better that it should proceed direct to the adult adaptations by the shortest possible route. Yet, while maintaining that the scientific problem is to ask how development *does* proceed, not how it *might* proceed, it may be said that it is, indeed, directly in the way of meeting such a theoretical criticism that we find all the abbreviations, 'short-cuts,' omitted stages, etc., which individual embryos actually show—the adaptations away from exact recapitulation of which recent discussions have made much.[1] The present writer has suggested[2] that, like everything else, the development of the individual must be subject to variations and such variations would be in turn subject to natural selection. Natural selection would operate wherever the recapitulation process was not the best

---

[1] For example, those of Sedgwick, *Quart. Journ. of Microscop. Science,* April, 1894, and Cunningham, *Science Progress,* I., N.S., p. 483.

[2] *Mental Development,* 1st ed., p. 32; see also *Dict. of Philos. and Psychol.,* art. 'Recapitulation.'

method of development, in a way to modify that process so soon as variations arose in lines of greater utility. This again has greater emphasis and stronger force in the light of intergenetic concurrence; a point which, in the writer's opinion, casts light upon the whole question of the relation of development and evolution to one another. It may be put as follows — as a second point in this discussion.

Second, when individuals acquire essential accommodations and modifications, these in so far mean the eradication or subordination of congenital characters which stand in their way or oppose them. If, then, concurrent evolution is to follow, it will be by variations which include the essential neutralization or cancellation of such characters. Succeeding generations must, if this principle holds, depart in these respects from strict recapitulation in all cases in which the environment does not allow to the individual all the stages in succession. Natural selection will seize upon the individuals which vary concurrently,[1] that is, in the direction of the *abbreviations and modifications of the genetic processes which are marked out first in the preceding generations' ontogeny*.

From this certain general consequences flow, each of which is illustrated by a large class of facts.

(1) A generation of a species may exist in an early simple form, an ancestral form, requiring little special adaptation; and afterwards, by some special mode of protection during later development, come unto the higher adaptations of the later forms of the phyletic series. So in the metamorphosis of many insects; the larva or worm

[1] Using the phrase to include the group of coincident and correlated variations, of which organic selection, as shown in the following pages, may make use.

stage of independent life being succeeded by the pupa or chrysalis, which has a protected mode of life in which the special adaptations of the later and more complex existence are made ready.[1] This is a case in which the evolution process has maintained the ancestral worm-stage intact.

(2) In the development of eggs deposited with shells, etc., and of uterine embryos, we find a device by which the early stages are accomplished under special protection without the independent early life seen in the cases of species having larvæ.[2] In the uterus all the environmental conditions necessary to development are realized as in the actual environment of ancestral forms, yet with varying detail; so that the internal uterine development is most favourable for the exhibition of recapitulation.

It is evident that embryological and such other modes of life as that illustrated in the chrysalis have thus their utility and 'reason for being'; for without them the preservation of the mode of development leading up to full heredity, as we find it, would be impossible, and so would, as a consequence, the special evolution of this or that species to its complex stage. The essential combination all along has been the accomplishment of progressive heredity with the addition of new adaptations. Recapitulation gives us a view of a former of these factors, the mechanism of heredity; concurrence shows us the oper-

---

[1] As this is being written, the extraordinary development of the seventeen-year locusts is going on, thousands of these creatures coming, in the writer's garden, from the earth, where their complete preparation for life has been made.

[2] See the experimental proof that the shell of the egg is such a protection to the development processes, by Weldon (research on the effect of introducing water, which, by promoting evaporation, hindered the development of the amnion), in *Biometrica*, I. 3, p. 368.

ation of the latter — which includes the very essential departures from recapitulation so often found in nature.

(3) In very simple organisms, which are known as 'generalized' as opposed to 'specialized,' and also in late stages of the development of higher animals, we find least evidence of recapitulation. In the simple organisms, heredity is still relatively unorganized, and the developmental series is shorter and more direct. In higher animals the periods of development which ensue after birth bring the animal into its independent life, in which its own adjustments to the environment are of capital importance. Hence no stages representing ancestral characters are preserved, except those which can exist in this separate life.

This last case seems to find illustration in mental development. We find that the series of stages of mental development does not show exact recapitulation; but that omissions occur.[1] In these cases variations have been rigidly selected in the line of intellectual endowment and educability, carrying with them increasing plasticity and lessened fixity, in the nervous substance and its connections. This is in line with some of the great correlations spoken of on an earlier page.[2] With it goes also the evolution of gregarious habits, family instincts, etc., by which the endowment of the infancy period in the highest animals is directly supplemented.

Third, we have another reason for the fact of recapitulation: the adaptations of hereditary endowment represent most essential adjustments to environment. Nature reaches them only after extended experimentation and great loss of life; really by the process of trial and error. It would

[1] See the cases cited in *Mental Development*, Chap. I. (especially the theory of 'short-cuts' in § 4).   [2] Chap. II. §§ 2, 3 (esp. p. 27).

seem, therefore, wise—speaking anthropomorphically—to preserve these essential adaptations and give each new generation the advantage of using them as so much capital or stock in trade. This sort of conservation the process of recapitulation reveals. To expose each generation to the chances of getting all their adaptations by chance variation with individual accommodation would be most disastrous to life; hence the preservation of the great lines of ancestral adaptation in each specific case.

The part played by individual accommodation appears, when the subject is looked at from such a point of view, in the successive cases of the development of the hereditary impulse in generation after generation. As pointed out by the 'organicist' writers, and as maintained in this work, an essential coöperation between heredity and accommodation is actually shown in the development of every individual that grows to maturity. There is thus a coöperation in development followed by concurrence in evolution, and the conservation of hereditary characters has to run the gantlet of natural selection under this essential coöperation in each case of life history. If, therefore, heredity, as revealing recapitulation, is a conserving engine, in the sense explained, then the decision as to how far it shall go and what details shall be conserved, really rests with the process of accommodation, in that it alone brings both hereditary characters and also new variations to their fruitful maturity.[1]

---

[1] The writer may say that the points made in this last paragraph (§ 5) are so essentially biological, that he states them merely as general inferences from the theory of orthoplasy, which nothing in his limited information contradicts, and which he has not come upon in the literature. So far as true they may be truisms, and so far as not truisms they may be not true, to professed experts in zoölogy.

CHAPTER XIV

§ 1. *Correlated Variations*

IN the preceding pages I have neglected, except by implication, the topic of the correlation of variations and of characters, refraining from asking the question of the locus of utility in the various spheres in which organic selection is supposed to be a real influence. It must have occurred to the reader, however, to ask whether the principle is not limited in its application to those modifications which confer direct utility upon certain variations. For it may be said that unless a given variation be made of direct utility by modification or accommodation, it would not be saved, nor would its possessor propagate his kind and so perpetuate such variations.

This is a pertinent inquiry; it indicates the limitation of the principle in all cases in which the perpetuation of a single variation of definite type depends upon the value of *modification of the same sort*. But just at this point the question of correlation comes in. A creature may be kept alive — and with him a great variety of characters may be perpetuated — through his accommodation and modification in some respect which may seem to have no bearing upon the particular character whose origin we are concerned to investigate. Creatures with better breathing capacity survive, and with them may possibly survive the tendency to have warts on the nose ; the wart-character would thus be preserved, although it may not have direct utility.

Indeed, many writers, as I have already pointed out, have recognized the facts which show hidden physiological correlations between things as apparently remote from each other as breathing capacity and warts on the nose. Under the operation of natural selection, variations in bony structure have to be correlated with variations in the muscles which are attached to the bones. A newly appearing character, which is as yet quite insignificant as regards utility, — such as the small lumps on the bone which the paleontologists fondle with such pleasure, as showing the first beginning of later developments of horn, or antler, or weapon of defence — such insignificant characters may advance *pari passu* with some remote modification, or be incidentally supplemented by some accommodation which is of utility and which thus acts as a screen to the former in the sense which organic selection postulates.

This principle — that one character may get the advantage of the utility of another and thus owe its permanence and development to it — has been recently and forcibly set forth by Professor Ray Lankester,[1] in a criticism of Professor Weldon, although the point was not new to the literature. The additional implication which I now note is that this holds not alone under the ordinary action of natural selection, where both the correlated characters — the valuable as well as the valueless one — are

[1] *Nature*, July 16, 1896, containing instances and citations from Darwin. Darwin's treatment of the subject is to be found in his *Variation in Animals and Plants*, Chap. XXV.

congenital variations; but also where the valuable and
life-preserving character is a modification, or an accommodation, acquired in the individual's lifetime, and serving
its purpose in connection with some quite different and
remote function.

We may use an illustration cited with all due emphasis
and triumph by Professor Cope in his *Factors of Organic
Evolution*, and accepted by an able critic, Mr. F. A.
Bather,[1] as affording evidence of Lamarckian inheritance
(provided the facts are true as reported). The appreciation of the case by the latter writer gives it additional
interest. The facts should be more fully inquired into,
however, with the sharp criticism which Weismann has
taught us to bring to bear on such cases.

The reported facts are: first, sheep carried from Ohio
to Texas produce wool which is harsh, when, before this,
the same sheep had wool which was fine and good; the
wool acts differently under dyes. This, it is suggested, is
due to the alkaline quality of the soil in Texas. So far
there is no difficulty; all would admit that the difference
is due to the conditions of the new environment upon the
individual sheep. Second, 'it is stated' (these are the
original reporter's words) 'that the acquired harshness
grows more pronounced with the successive shearings
of the same sheep;' this again may be true and due to
the continued direct action of the environment. So far,
no trouble. Third, 'it is also alleged' (again the reporter's words, together with what follows in quotation
marks) 'that the harshness increases with succeeding
generations, and that the flocks which have inhabited such
regions for several generations produce naturally a harsher

---

[1] *Natural Science*, January, 1897, pp. 37 f.

wool than did their ancestors, or do the newcomers.'
Now let us assume that what is ' alleged ' is true; still
we find that it is not said that there is any evidence that
the young sheep inherit the harshness of wool; it is only
alleged that the 'harshness increases with succeeding
generations,' and that 'the generations which have inhabited such regions for several generations produce naturally a harsher wool.' There is absolutely nothing here
to lead us to believe that the harshness has become congenital at all. Wool is not cut till the sheep has been
alive long enough to grow it. So it is natural to think
that each sheep acquires the harshness for himself. But
how account for the increasing harshness in succeeding
generations? That is what has impressed Mr. Bather.
This might appear true from the fact that each shearing
of the wool of the same sheep was harsher than the last;
for in order to compare the harshness of two generations,
sheep of the same age, measured by the number of times
they had been sheared, would have to be compared. But
again, waiving this, let us assume that there is a congenital
difference, — the quality of wool being more harsh for
later generations, — then how can we account for this
increased congenital tendency to harshness?

We might say that the increase in the harshness of the
wool in subsequent generations was due to the natural
selection of sheep with congenital variations; this would
be open to the objections so frequently urged against natural selection, that it is not likely that such variations
would come in such great numbers; and also to the objection that the difference in harshness of the wool might not
be of utility. But the principle of organic selection acting on correlated variations would meet both these objec-

tions; for we only have to say that there is some acquired physiological accommodation with which the harshness of the wool is associated or correlated; this physiological accommodation utilizes variations present in a greater or less number of sheep; these sheep survive by natural selection, and produce the next generation; the next and subsequent generations have further variations in the direction of this physiological adaptation; and with it will go the increased tendency to have harshness of wool. The rapidity with which this process would go on would depend upon the importance and the difficulty of the physiological adjustment. Say, for example, that the change in diet, soil, climate, etc., lays the imported sheep open to a certain disease; therefore, a certain strength, ruggedness, vigour of constitution, which carries with it harshness of wool, is necessary to give recovery and gradual immunity from this disease; then the sheep which lived would continue the variations which tended to permanent immunity, and with them would go the harshness of wool. That this is the case seems tolerably plain from the nature of the character, and from the fact that it increases from one shearing to another. Such a change in the quality of wool could not take place incidentally; it could not arise by selection without some specific utility; it must represent some deep-seated adjustment. The quickness with which it takes effect in the individual sheep would show the likelihood that its rapid individual acquisition was necessary to save the sheep. Put tersely, the sheep are saved by accommodation, and with them are saved both variations toward that accommodation and also other characters which are correlated with these.[1]

[1] It is necessary to note that the question whether these sheep retain the harshness of wool when taken back to Ohio is not answered by our informers.

§ 2. *The Coincident Variation Theory not Sufficient*

Just here there seems to be a point of difference [1] of construction of the principle of organic selection; or at least a difference of emphasis, which results in giving it somewhat different range. Professor Morgan defines the principle as, in effect, 'the natural selection of coincident variations,' [2] *i.e.*, variations in the same direction as the modifications by which they are shielded and with which they are said, for this reason, to 'coincide.' In view of the application of the principle pointed out above, the writer, on the contrary, includes not only 'coincident' but also *correlated* variations. It is indeed true that the accommodations and modifications, in so far as they are directly supplementary to an incomplete organ or function, open the way for coincident variations; these, gradually appearing in the direction of the modification, in time replace it. But accommodation in many cases — indeed, possibly in most cases — involves a complex mechanism, a complex group of characters. It keeps alive not only the variations which coincide with it in a particular function; it keeps alive the whole animal, and so screens all the characters which that animal has. And various lines of adaptation may be fostered and screened by a single general accommodation. This appears notably in the case of the intelligence. Intelligence is what is called above a 'blanket utility'; it comes into play again and again to supplement the most varied

If they did retain the harshness, then that would be an illustration of what I have indicated above: the same dependence on variations to get rid of the character that there was in acquiring it, *i.e.*, by reversed selection while organic accommodation was in operation.

[1] As between Professor Lloyd Morgan's views and the writer's; see, however, the 'new statement' made by Professor Morgan in Appendix A.

[2] *Animal Behaviour*, p. 115.

functions; and its influence may be diverse, according as this or that animal, differing from others in his correlations, finds himself able to use it. It has been said above that divergent lines of evolution may spring up *because fostered by one and the same accommodation.* This means that the division of an animal into set characters is at the best artificial; the whole animal lives or dies, and some slight utility here, or another slight utility there, will give natural selection its chance on this individual or that for the emphasis of this character or that. The correlated characters are just as truly screened and fostered, and helped over the hard places, by organic selection as are the coincident characters. And such is the plasticity of the organism that such characters may be pressed into service for utilities ' not dreamed of ' in the original function by which they may have been kept alive.

§ 3. *Illustrations of Orthoplasy with Correlated Variation*

Let us take as illustration the case of the evolution of the sole, the adaptation of whose eyes has already been remarked upon (p. 164). This adaptation is part of a larger and more complex one. Indeed, the placing of both eyes on the same side has utility only to an animal which lies on one side near the ground, and so does not require an eye on the under side; this position — flat on one side — is therefore the adaptation in connection with which the position of the eyes has its utility. But why is it useful to the sole to lie on one side near the bottom? — why this adaptation? This is explained again, in turn, by another adaptation, that of concealment from enemies, and for this utility we find another series of correlated changes, those of protective coloration — the under side.

of the sole becoming of a light colour which approximates the light colour of the water *when seen from beneath,* and the upper side a dull gray or mud colour, approximating the surroundings *when seen from above.*[1] It is probable, therefore, that all these striking adaptations serve the great and prime utility of concealment. If we now revert to the accommodation of the position of the eyes, we find what may be a striking illustration of the operation of organic selection in screening correlated characters.

For we may assume that so long as the adaptations in coloration, and especially in the position of the eyes, were not secured, or were only partially evolved, it would not be of utility for the sole to lie on the side, for the upper side would be exposed to view, and there would be the disability arising from rendering one eye useless, if the fish took such a position. But the coloration, in its turn, could not be acquired through natural selection so long as the fish did not lie on the side. Accordingly it seems a fair inference that *this whole group of adaptations required such a gradual adjustment of the eyes;* that the maintenance of the function of vision unimpaired was absolutely necessary to the sole if he was to escape elimination, while gradually, as the adjustments of the position of the eyes went on, he acquired variations placing him more and more on one side, and also variations in the direction of the requisite protective colouring. The straining of the

---

[1] Professor Osborn, who has kindly looked over the proofs of this chapter, suggests (citing Cunningham's experiments in evoking colour on the lower side of the sole, by throwing light upon it with a mirror; cf. Cunningham on 'Recapitulation,' in *Science Progress,* I., N.S., pp. 483 ff.) that the colour-differences are ontogenic, that is, that they are accommodations acquired in each generation. This fully accords, in so far, with the position taken below, that this adaptation is secondary to that of vision while the fish is in the flat position.

eye muscles in each generation allowed a more flattened position without impairment of vision ; this position was selected for its utility, particularly when correlated with variations in the direction of the protective colouring ; and with the accumulated variations in bodily position and in coloration, *thus screened by the eye adjustments*, the sole as we find him attained perfection.   This, at any rate, — whatever we may say as to the leading rôle thus assigned to vision in this series of adaptations, — shows how *a single accommodation of great utility may screen not only coincident variations, but also correlations of variations*, by which a species' characters and habits of life are in a remarkable way transformed.

In this case it is only the variations in the eye position that can in any sense be called 'coincident' with the modifications acquired by the individual soles by the use of their eyes ; but the other great systems of variations, in bodily position and in coloration, are just as truly screened and developed by these modifications.

Another instance may be cited, which is due to the writer's observation, and which is accordingly submitted with some misgiving to the scrutiny of expert ornithologists.

The gray parrot (*Psittacus erythacus*, Linn.) from the west of Africa has, in company with all individuals of his kind, a very strong upper mandible which is curved sharply downward at the extremity, so that its very sharp point is directed downwards about at right angles to the line of the whole beak.   It seems that this curving of the mandible downward subserves the bird great utility as a sort of third foot.   In getting down from his perch he constantly extends his head, rests the beak upon the bottom of the cage, and then alights on one foot, while holding to the perch as long as possible with the other. So constant and uniform an act is this — as also with the companion bird of the same species — that it may, I think, be considered a real and very useful adaptation. Further, the advantage of having this utility subserved in this way instead of by a blunt beak is, that the sharp point of the beak, while no longer a hinderance to such a use of the mandible, — as by penetrating the ground, catching in the texture of any material he may be resting upon, etc., — is, nevertheless, not lost; and it is of very great service in biting, rending, breaking nuts, etc.

Now, if we admit this utility — that of a sort of third foot — in the upper mandible, the question arises how the adaptation may have been acquired. In answer to this, we may cite the further character — in these parrots — that the upper mandible is somewhat loosely attached, the muscles allowing rather free movement, as is the case with many other birds; and it is in connection with the relative flexibility of the upper mandible, in its relation· to the head of the bird, that the curvature of the beak may have been acquired by a gradual operation of natural selection. If the mandible was at first straight, and if there were variations in the relative flexibility and range of the muscles by which it is attached, the use of the beak more or less clumsily for descending would be possible to some of the more flexible parrots. This muscular accommodation would screen variations in the direction of the curved beak; for, the more curved, the better could the function be performed. This process would continue from generation to generation, until the adaptation attained, by variation and natural selection, the degree of perfection it has in the present-day parrots.

Of course the availability of this case as an illustration of organic selection — apart from the facts — would depend upon whether the flexible attachment of the upper mandible did supplement the curvature of the bill, or take the place of it, in the early stages of the adaptation. This appears the more likely from the further fact that this same character, the relatively loose attachment and somewhat free muscular play of the upper mandible, is useful for other purposes as well. In feeding, the parrot makes much use of these muscles, using his mandibles as a nut-cracker, and in general for crushing hard objects between them. It also comes in 'handy,' so to speak, in a variety of those gymnastic feats whereby he exercises his agility in a manner truly serpentine, using his beak as a claw for holding on and lifting himself. In all this the flexibility of the upper mandible is of direct advantage.[1] It is also well known that the vibration of the upper mandible greatly enriches the range and variety of bird song, and in this case it may be of great utility in the parrot's extraordinary power of articulation.

If this be a true construction of the facts of the case, it will at once be seen to illustrate the correlation of variations which are now referred to.[2] For not only do we find variations of a sort which we may call coincident, — in this case variations in the flexibility of the muscles whereby greater movement and control are possible, — but also in the shape of the beak. And if the other utilities spoken of be real, possibly also variations in the claws and

[1] A very comical exhibition of the amount the parrot can raise the upper mandible is seen when he *yawns*.

[2] It illustrates the method of such correlation in any case.

vocal organs are correlated with the loosely attached upper mandible.

Again, suppose we take such a generalized function as imitation. It is not shut down to a single particular channel of expression, but applies equally to a great variety of alternative and possible functions which the imitative creature may be at any time led to exercise. This tendency to imitate screens many incomplete functions, and thus serves very essential utilities in life history. But this does not mean that in each case variations are fostered only toward the congenital performance of the function in just the same way that the imitative performance secured it. On the contrary, as is said on an earlier page, and as has been maintained by Professor Groos, the imitative functions act in many cases to develop the intelligence, to enable the creatures to do many things by a type of action which discourages evolution by coincident variation in these directions, and encourages variation of the opposite sort, that, namely, toward greater plasticity and intelligence. Here the saving utility of individual accommodation is exercised in the way postulated by organic selection : it screens the organism and utilizes a partially congenital and quasi-reflex mechanism ; but its racial utility in connection with many functions seems to be just the reverse of the selection of coincident variations. Yet in other cases, indeed, as is maintained in an earlier discussion of the subject (Chap. VI., above), its results are strictly in accord with the theory of the development of coincident variations.

Apparently — in view of these illustrations and others which might be cited[1] — we may look to the accommodations

---

[1] As, for example, that of the evolution of mammalian teeth, mentioned by Professor Osborn (quoted in Appendix A, I.). If, as is suggested in App. A, I.,

of the organism as having very far-reaching and often un-expected utilities. An adjustment, whose immediate utility is that it supplements and screens incipient characters, may by a slight change in the environment, or by support-ing variations in a neighbouring part, or by another adjust-ment of other organs, become part of a new system of adaptations not at first accomplished by it at all; there may arise from the countless variations in shape, size, and relation of parts, utilities that no one could have predicted, and which only natural selection can and does discover; the very flexibility upon which the principle of organic selection lays emphasis tends to reduce our expectation that single lines of characters, coincident or other, will appear, for by it all sorts of alternative and shifting utili-ties are allowed to spring up.

The limitation of the application of the theory to coin-cident variations would therefore, in the present writer's opinion, serve to take from it much of its value, that is, if 'coincident' be defined strictly, as it seems to be by Professor Lloyd Morgan. If, however, we attempt to bring all the phenomena under this term, it loses much of its appropriateness; for it would have to be defined to include all functions and characters which might evolve in the whole organism, in consequence of a particular accom-modation, and *become substitutes for the accommodation, or in any way replace it.* Accordingly, while the theory of coincident variations is true and covers much of the terri-tory, furnishing most valuable illustrations of the working

muscular adjustments compensate for wear and tear on the teeth in the individual's life, the evolution of the teeth, although screened by these ad-justments, would nevertheless be by variations directly *contrary* to the modi-fications wrought by their use.

of organic selection, yet the latter, in my opinion, can nevertheless not be defined, as it has been (cf. the *Année Biologique*, V., 1901, p. 388), as 'the natural selection of coincident variations.' And the criticisms, as, for example, that reported (*ibid.*, p. 388) as made by Plate, which get their point from the limitation of the application of the theory to cases of coincident variations, lose their force when we recognize that such a limitation is not necessary.

§ 4. *Natural Selection still Necessary*

Such a criticism takes the form of the question as to the further utility of congenital variations, especially those of the coincident tendency, when by the use of accommodations, the individuals can already cope with the environment. Put generally, this criticism would read: does not the theory of organic selection, by showing that accommodation does supplement imperfect organs and functions, make it unnecessary that variation and natural selection should be further operative? This leads, it would appear, to the extreme position of the organicists, as is illustrated by the quotation made from Pfeffer on an earlier page.[1] It seems to be also the opinion of Delage (see the *Année Biologique*, III., 1899, p. 512).

This criticism is fully met, I think, when we remember that natural selection may seize upon any utility, additional to that already springing from any functions which animals may perform, no matter how they may perform them. Many functions may be passably performed through accommodation, supplementing congenital characters, which would be better performed were the congenital characters strengthened. Congenital variation would in these cases

[1] Note to p. 184.

P

by seizing upon this additional utility, carry evolution on farther than it had gone before. For example, muscular strength in biting would in no way prevent the evolution of hardness of teeth. The accommodation factor would be gradually dispensed with, since the most unsuccessful of those which depended upon accommodation would be eliminated. In the case of an instinct, for example, which represents congenital endowment at its best, this would give the gradual shifting of the congenital mean toward the full endowment, even though the creatures could — or some of them could — still survive on the earlier basis of strenuous accommodation. It would be a case — as in all other cases of natural selection — of more or fewer individuals surviving, and a consequent shifting of the mean. Moreover, as is pointed out in an earlier place,[1] in many instances we find both types of function, the congenital and the accommodative, serving somewhat different utilities, and so existing together.

At the same time we have the advantage of recognizing the state of things which the organicists point out, in cases in which it exists. There are undoubtedly functions for which the accumulation of congenital variations would have no utility, or would be of positive disutility. In these cases we find either a state of 'balance' between the organism and its environment, or the actual decay of congenital functions. In the state of balance, the accommodations of the individuals would be made again and again in successive generations, and no further development of congenital endowment would take place. This flexibility of application which the principle of organic selection allows, seems to be one of its great advantages.

[1] Chapter VI. § 1.

There are positive grounds, indeed, — to take the matter further, — for discarding the extreme position of those who deny the part in evolution played by congenital variations accumulated by natural selection.   The specific character, the persistence, and the definiteness of the hereditary impulse, require that we should recognize its leading rôle in development, despite the large part attributed to individual accommodation.   All the evidence accumulated by writers since Darwin in support of natural selection operating upon variations, together with the statistical work upon variations, is available to show that heredity represents a real and definite impulse which conserves the specific type, and in large measure the specific characters, of organisms.   Recent work in experimental morphology emphasizes the persistence of heredity in reverting back to its normal development in the individual, as soon as artificial conditions under which it may have been modified are removed.[1]   This persistence appears in the life history of twins, in the phenomena of atavism, in 'exclusive' as opposed to 'blended' heredity,[2] in the protected development of pupæ, etc.

This granted, variations in congenital endowment at once become liable to the operation of natural selection for any utility they may serve, and this the more when they are supplemented by individual accommodations.

An additional general remark is suggested by this criticism.   It should always be borne in mind that a theory of evolution does not attempt to account for organs nor

[1] This has been dwelt upon by Professor E. B. Wilson.

[2] On the persistence of hereditary traits in twins see Galton, *Enquiries into Human Faculty*, pp. 216 ff.   On 'exclusive' Heredity (*i.e.*, heredity from one parent only) see Galton, *Natural Inheritance*, p. 12 (cf. also the Index to that work).

characters which do not actually exist, not does it attempt to say that such or such a thing might exist. On the contrary, it simply aims to show that such or such an actually existing structure, mode of behaviour, etc., has probably arisen through the operation of the forces and principles which the theory recognizes. The actually existing forms are so varied that different emphasis must be placed, now on one factor of the whole process, now on another. Instinct, for example, seems to require, for any explanation approaching adequacy, the factor of accommodation to supplement that of natural selection. Mimicry and those anatomical and structural characters in which the element of function is much reduced, seem to be explained by natural selection with little supplementing from other factors. It may be found ultimately that Lamarckian heredity holds for simple organisms and for plants, while in higher organisms and in mammals it is not operative. The problem in each case, therefore, may be stated thus: the fact is that such an organ exists; its utility can be explained only on the supposition that accommodation coöperated with any congenital variations which may have existed; it has thus evolved up to the stage which it actually shows — complete function, partial function, mere beginning, as the case may be; it is quite possible, had the conditions favoured it, that its evolution might have gone farther, or, indeed, not so far; but that it did go so far, and no farther, is in itself sufficient evidence of the utility of the coöperation of heredity and accommodation in its production.[1]

[1] See the insistence on Natural Selection in Professor Ll. Morgan's 'New Statement,' in Appendix A.

PART III

CHAPTER XV

STRUGGLE FOR EXISTENCE AND RIVALRY

§ 1. *Biological Struggle for Existence*

THE struggle for existence may be defined as the attempt to remain alive, or technically to 'survive,' on the part of an organism. As a necessary factor in Darwinism, the conception involves the further restrictions, which, however, are not so generally made clear: (1) that the organism which survives is already, or is still, capable of propagating in the manner normal to its species; and (2) that it finds opportunity to do so; failing either of these conditions, the case would not be one of successful struggle for existence, from the point of view of the theory of descent.[1]

Three clearly distinguishable forms of struggle for existence may be recognized:[2] —

---

[1] Darwin says (quoted by Bosanquet): I use this term (struggle for existence) in a large and metaphorical sense, including dependence of one being on another, and including, what is more important, not only the life of the individual, but success in leaving progeny.

[2] Cf. the writer's *Dictionary of Philosophy and Psychology*, arts. 'Existence (Struggle for)' and 'Rivalry,' where the following distinctions are made out. Distinctions among different forms of struggle are made by Pearson in *The Grammar of Science*, 2d ed., p. 364, who distinguishes 'struggle of individual man against individual man, struggle of individual society against indi-

(1) The competition for food, etc., that arises among organic beings from the overproduction of individuals or from a limited supply of food. This is called the 'Malthusian form' of struggle for existence.[1]

(2) Competition in any form of active contest in which individuals are pitted against one another.

(3) Survival due to greater fitness for life in a given environment, whether combined with direct competition with other organisms or not.

The second case (2) is that in which animals either (*a*) fight with, or (*b*) prey upon, one another; only the former of these (*a*) having any analogy to the form of competition due to the overproduction of individuals or to a limited supply of food, etc., and then only in the case in which the strife results from the circumstances of getting a living — not in the case very common in nature of mere combativeness of temper, through which the stronger animal kills the weaker simply from aggressiveness. In case (*b*) one animal feeds upon members of another group as his natural prey, as is seen in the eating of insects by birds. This is also extremely widespread, and leads to some of the most beautiful special adaptations — for concealment, warning, etc. — in the species preyed upon. This has nothing to do with the overproduction of individuals in the sense given under (1), except in so far as the species preyed upon overproduces in the way of compensation for the constant drain upon it; but this is a very different thing.

vidual society, struggle of the totality of humanity with its organic and inorganic environment.'   See Chapter XIX, § 7, for a criticism of certain of Professor Pearson's positions.

[1] From the fact that both Darwin and Wallace were indebted to Malthus' work *On Population* (see below).

A case of (2*a*), of extreme importance in its effects upon the next generation, is that of the struggle of males for the female, occurring often apparently irrespective of the number of available females.

The third case of struggle (3) is that in which individuals struggle against fate — the inorganic environment — not directly against one another. This is really a 'struggle to accommodate' — to reach a state of adjustment or balance under which continued living is possible. As the other forms may perhaps be styled respectively 'struggle to eat' (in a large sense) and 'struggle to win,' so this may be called 'struggle to accommodate,' or 'struggle to live.' The distinction between cases (2) and (3) disappears in instances in which the animal accommodates actively in order to coping with his enemies; for these then become part of his environment in the sense of case (3).

The relation of large productiveness to this last form (3) of the struggle for existence would seem to be but indirect. It would not matter how many individuals perished provided some lived; and any amount of overproduction would not help matters if none of the individuals could cope with the environment. Yet on the theory of indeterminate or indefinite variation, the chances — under the law of probability — of the occurrence of any required variation is a definite quantity, and these chances are of course increased with large productiveness; for with more variations, more chances of those that are fit, and with increased production, more variations. No better case in point could be cited than Dallinger's experiments on the effects of changes of temperature on infusoria.[1]

In recent evolution theory the doctrine of natural selec-

[1] An illustration suggested by Professor E. B. Poulton.

tion has come to rest more and more upon the second and third sorts of struggle (2 and 3), and less on the Malthusian conception (1). Experimental studies which support the selection view (*e.g.*, Weldon on Crabs, Poulton on Chrysalides)[1] show the eliminative effect of the environment, and the preying of some animals upon others, rather than direct competition *inter se*, among individuals of the same species, for food or other necessities of life. It is these forms of the struggle, too, that we find nature especially providing to meet, through adaptive contrivances such as concealing and warning colours, mimicry, offensive and defensive organs — teeth, claws, horns, etc., — with combative, aggressive, and predatory instincts, on the one hand, and by high plasticity and intelligence on the other.

The result common to all the sorts of struggle for existence, however, is the survival of an adequate number of the fittest individuals; and this justifies the use of the term in the theory of evolution to cover so wide a variety of instances.

Darwin, on reading Malthus' book *On Population*, conceived the idea that overpopulation would be a universal fact in organic nature were there no process by which the numbers were constantly reduced. He was thus led to lay stress on the struggle for existence, and the elimination of those individuals which were unsuccessful. Combining this conception of elimination in the struggle with that of variation, he reached the hypothesis of natural selection. A similar relation to Malthus is also true of Wallace (see Poulton, *Charles Darwin*, pp. 88 f.)[2] In this case the struggle arises from common wants, combined with

---

[1] Weldon, *Proc. Roy. Soc. London*, LVII., pp. 360 ff., 379 ff.; Poulton, *Proc. Brit. Ass.*, Bristol Meeting, 1898.

[2] See also p. 46 of the same work.

an inadequate supply, the competition taking the form either of direct struggle of one animal with another, or death from mere lack of something necessary on the part of some. This conception has been broadened with the development of the theory to include the other less Malthusian forms.

If we consider the three forms of struggle pointed out above, as together making up the conception, we may for convenience designate it as 'biological' struggle, inasmuch as individuals are directly brought into conflict with one another for life and death, and as moreover the end is not attained through the struggle alone, but requires the further biological function of reproduction to make it effective.

In greater or less contrast with this, we find other cases in which there is a shading one way or the other away from this form of competition with its indirect results. On the one hand, there are certain hypotheses of a biological sort which utilize the conception of struggle without distinguishing it clearly as a process preliminary to that of survival. In Roux' 'struggle of the parts' the conception is of the relative determination of physiological processes by the accentuation or development of certain cells and organs at the expense of others.[1] It is analogous to the struggle for food; the idea being that there is a preferential supply of nourishment, blood — whatever aids the anabolic processes in these particular directions. But the mechanism of it is entirely unknown. In Weismann's 'germinal selection,' also, a similar reason for survival is postulated — differences of some sort between germ-cells — whereby some of them are more favourably situated or otherwise

---

[1] Roux, *Gesammelte abhandlungen über Entwicklungsmechanik der Organismen*, Vol. I.

conditioned for survival, and this may be called a 'struggle.'
But here again there is no precise notion of what takes
place.   In both of these cases the struggle is merely a
hypothetical means to the end, which is selection and survival; it is not a clearly described phase in the process.

Weismann's 'Intra-selection' also involves struggle, in
an obscure way; and the selection of motor functions
by what has been above called 'functional selection' involves the survival of movements from among a series of
overproduced or excessive discharges; but it again seems
to strain the notion of 'struggle for existence' to speak of
these movement variations as engaged in a struggle with
one another or with the environment.   In all these cases
Mr. Spencer's term 'survival of the fittest' is more applicable; and the criteria of utility and adaptation run through
them all.

§ 2. *Sorts of Rivalry*

Coming to the extensions of meaning of the concept of
survival with struggle, in the direction of conscious and
social functions, we find certain processes which we may
distinguish under the general heading of 'Rivalry' — a
broad term which may be used to designate the entire
field, including biological struggle for existence.

There are three great cases of Rivalry which it is essential to distinguish, especially in view of current confusions
arising from lack of discrimination: (1) Biological Rivalry,
or struggle for existence, of which the forms have been
pointed out above; (2) Personal or Conscious Rivalry, or
emulation, to which the term rivalry is more generally
restricted; and (3) commercial and industrial rivalry,
known as Economic Competition.

§ 3. *Conscious Rivalry*

The second of these, personal conscious rivalry, is the relation between two persons, or more, which arises from their mutual intention or effort to excel each other in attaining an end which they have in common. It is distinguished from biological struggle by two marks, at least.

In the first place, it is for the sake of a further or remote conscious end that this form of rivalry is usually indulged in; the competition itself is a means to another end. There may be cases, indeed, notably in autonomic functions such as play, in which no end apart from the function itself is set up; but even in these cases the element of rivalry — as in the contests of a boy's game — is an incident of the game, not a thing indulged in for its own sake. And even in the extreme case of games of rivalry as such, in which the competition is the main motive, the fact of its being play destroys its analogy to struggle for existence in the biological sense.

A second difference is in respect to the immediateness or mediateness of the results. As pointed out above, struggle for existence is really biologically effective only if reproduction and physical heredity ensue to clinch and further the results of the struggle. If the individuals which remain do not produce young, they have not survived *biologically*. So the effectiveness of struggle for existence is secured only through the medium of the further vital function of reproduction. In personal rivalry, on the contrary, this is not the case. The results are immediate. The rivalry furthers the end for which the conscious competition takes place.

In personal rivalry, in fact, we have all forms of individual competition for personal pleasure, profit, gain, victory, etc. It is, so far as the actual contest goes, similar to the second form of struggle for existence; but it is narrower, since it includes only those cases in which the individuals are directly and consciously exerting themselves against each other. It is, therefore, always a psychological fact, as a little further analysis will show.

The psychological factors involved include: (*a*) the particular impulse appealed to to excite the effort — whether 'desire of being a cause' (Groos), called in the older English literature 'love of power'; desire to gain advantage, — pleasure, reward, gratified pride, etc., — earlier designated 'love of gain'; intellectual exercise — play of the faculties; or other. Any or all of these enter in cases of personal rivalry; and in adult life probably there are also in many instances reflective motives, such as pure love of success, love of the game as such, malice toward competitors and jealousy of them. (*b*) The psychological requisites of the *personal-rivalry situation* as a whole. These are those of the social bond, in which the self and the other (*ego* and *alter*) are held in a common network of social relationships within which the contest takes place, and by which its rules and conditions are prescribed. This, it is well to note, involves as much coöperation as competition. The rivalry is never entirely rivalry, and it could not be rivalry at all, in the more complex cases, but for the great mass of coöperative thinking, feeling, and action which precedes and conditions it. In short, personal rivalry involves an essentially coöperative factor; it implies a social situation in which, it is true, the pole of self-emphasis, assertion, and even aggression is very prominent, but in which, nevertheless, that is only one pole of the play of elements which constitute the thought of self as a 'socius' or personal companion to others.

Personal rivalry is, therefore, sharply distinguished from biological struggle for existence. The latter is operative under the law of physical reproduction, guided by natural selection with reference to utility in a biological environment. This, on the contrary, is operative in a social environment where social tradition through imitation and invention are the conserving and ordering factors, where the environment is psychological and moral, and where the criterion of utility yields to that of individual choice, selection, reflection, and, it may be, caprice.

This is not to say, however, that personal rivalry may not be involved in biological survival. It is evident that the capacity for personal psychologically motived struggle may be of critical utility to a species, and so its possessors may be 'naturally' selected. But true as that is, such a case still remains one of biological struggle, and is subject to its laws. The purely social rivalry, as such, remains a different phenomenon, and cannot be subsumed under the biological.

§ 4. *Economic Rivalry or Competition*

In industrial and commercial competition we find another form of Rivalry — that mentioned third above. It is defined by Hadley (the writer's *Dict. of Philos. and Psychol.*) as 'the effort of different individuals engaged in the same line of activity each to benefit himself, generally at the others' expense, by rendering increased service to outside parties.'

Two typical forms of it should be distinguished: (*a*) competition of individuals, which we may call 'free' competition; and (2) competition of agencies, either individuals or organizations, which we may call 'restricted' competition. This distinction is essential, for it indicates two types of competitive activity.

(*a*) Free competition, considered as a type operative in industrial and commercial affairs, leaves to the individual, in his attempt to succeed, freedom of enterprise, initiative, and method of operation. It is, therefore, psychologically motived, and rests directly upon the individual's capacity, temperament, and social feeling. The economic motive is tempered and modified by the individual's character, and varies all the way from pure egoism or love of gain to the most humane and social concern for others' welfare and success. It appears, therefore, that in free competition we have in operation the factors involved in personal rivalry, but directed to an economic end. This end in view gives to the agencies of production, trade, etc., a certain real aloofness which appears inhuman, and is often made the excuse for what is really so; but yet industrial organization, in which free competition is the dominant form, is a mode of social organization in which the factors involved are those essential to the maintenance of social life, and consistent with its other and more altruistic modes. Hence the growth, within the ordinary machinery of industrial economics, of various purely social and ethical features — humane labour laws, hygienic surroundings, libraries and reading-rooms, baths, lecture courses, lyceums, etc., not only permitted but provided by employers, together with such more intrinsic arrangements as profit-sharing, increasing wage, pensions, labour insurance, etc. In essentials, therefore, this form of competition does not merely represent but *is* personal rivalry inside the industrial world.   It is not, *nor is it analogous to*, biological struggle for existence.

In another point commercial competition involves psychological factors; appeal is made to the desire and choice of the consumer — what is known as 'demand.'   This demand may be reached either by direct rivalry for the consumer's patronage, or indirectly through means which increase the use of certain articles, set the style, limit variety, etc.   In these ways of directing, stimulating, and controlling demand, *all the competitors may be alike benefited by the success of one.*   This is different from the use of brute force, and also from the division of a fixed amount of patronage or gain — processes which would present analogies with the usual methods of biological rivalry.

(*b*) The second form of economic rivalry — 'restricted' competition — is a different matter.   It arises when individuals band together either voluntarily or under social compulsion or persuasion to pursue common economic ends in association.   This gives to the group economic standing *as an agency;* and the members cease to act as individuals.   The result is the formulation of purely economic rules of procedure — of defence and offence — and the elimination of individual temper, judgment, and sense of personal and social responsibility.

The direct result is that such a society becomes a *group*, and when engaged in competition with other groups gives the phenomenon of 'group selection,' yet it is group selection, in the strict biological sense, only in part.   As to the struggle, strictly speaking, of group with group — it is struggle for existence in so far as it means elimination of some groups and survival of others.   But its results

are socially conserved and handed down, not passed on by
physical reproduction and heredity.   So the resemblance
is still in part analogy.   Even restricted competition is
not a biological fact; in its most ironclad and 'inhuman'
forms it is intelligent; intelligence unmoved by feeling
is its watchword.   And its forms of rivalry are very
largely those of one master intelligence pitted against
another.

Yet in this phenomenon of restricted competition we
have the nearest social approach to biological rivalry as
such; and that in certain unæsthetic features in which
economic utility is the controlling end, if not the only
one.   First among these is the opportunity it affords of
subordinating and destroying normal personal competition with its natural control by social and moral sentiment.   Second, there follows, the need of state control
to take the place of other controls; there would seem to
be no other alternative.   Third, we find not only group
competing with group, but class organizations arrayed
against each other, when the closest coöperation is essential even for the purest economical utilities; as of labour
organizations against capital, employer against employee.
And fourth, all are contributory to the great damage done
to society by the interference with personal liberty of
contract and choice of work under the oppressive sanc-
tions of the organizations, which claim to regulate economic
conditions.   In all these respects the industrial environment in which modern corporate agencies operate is
analogous to the biological; for utility is the criterion of
survival, and economic utility is in many respects analogous to biological.

The contrast presented by the three great sorts of rivalry

now distinguished are sharply brought out when we ask what forms of coöperation of individuals they severally involve. So far as coöperation enters into biological struggle for existence, it is instinctive and unreflective — as in the gregarious and mass-actions of herds or other companies of animals. It is a phenomenon of a biological sort produced by the operation of natural selection. Intelligent coöperation, to be available in the struggle, has utility not in the direct results of the coöperation, but as representing a type of individual which it is of utility to preserve by the laws of heredity.

In personal rivalry, and with it free economic competition, we have the intelligent and reflective coöperation which illustrates the presence of a social and moral self in some degree of development.

In 'restricted' competition we revert to an economic formula which makes utility paramount, and only that form of coöperation possible which subserves this utility. This may arise among individuals within the group so far as it renders the group as such more efficient as against others — and also as between different groups or agencies for the ends of common utility.

Q

CHAPTER XVI

LAMARCKIAN HEREDITY AND TELEOLOGY

§ 1. *The Evidence in favour of Use-inheritance*

THE evidence for the inheritance of acquired characters, called 'Lamarckian' or 'use-inheritance,' in cases of sexual reproduction, is not very strong. There are no clear and unambiguous cases of transmission of specific modifications. The arguments for such transmission are largely presumptive, based upon the requirements of the theory of evolution. Of such arguments the following seem to be the strongest.

(1) Incomplete or imperfect instincts—together with complex instincts, which must at some time have been imperfect—cannot be due to natural selection; for their early stages would involve partial correlations of movement of no use to the animal. Selectionists meet this by saying that (*a*) the organism as a whole must be considered, not the single organs or functions, in the matter of individual survival; (*b*) a certain degree of intelligence usually accompanies and supplements such instincts; (*c*) the intelligence, together with individual accommodations of all sorts, screens those variations which occur in the direction of the particular function, and secures its evolution under natural selection in accordance with the hypothesis of *organic selection;* (*d*) many of the instances cited under this head are not congenital characters at all, but are

226

functions reacquired in whole or part by the young of each succeeding generation.

The Lamarckians urge (2) that paleontologists find bony structures whose initial and early stages are thought to have had no utility; and appeal is made generally to the so-called non-useful stages of useful organs. This is conceded by many to be the gravest objection now current to the universal applicability of natural selection.[1] It is met — when urged as giving presumptive evidence of the transmission of acquired characters — by saying : (*a*) that it proves too much; for the bones are of all the structures least subject to modification by external influences, and if such inheritance appears in them, it should appear more strongly in other structures where we do not find evidence of it; (*b*) that even if such an objection should be found to hold against natural selection, still some unknown auxiliary factor may be operative; (*c*) that actual utility can be pointed out in most cases, and may be fairly assumed in others; (*d*) organic or indirect selection again has application here, as supplementary to natural selection; (*e*) the principle of 'change of function' (*Functionswechsel;* see A. Dohrn, *Der Ursprung der Wirbeltiere und das Princip des Functionswechsels*, 1875) is cited, according to which, in such 'non-useful' stages, the organ in question served another useful function and was selected for this utility.

Other arguments are mainly negative, consisting largely of objections of a general sort to the sufficiency of natural selection — such as that geological time is not sufficient for so slow a process as evolution by natural selection, that small variations could not produce such large aggre-

[1] Cf. Chap. V. § 3, and Chap. X. §§ 1, 2.

gate differences, that variations are not sufficiently numerous nor sufficiently wide in distribution. These are considered by selectionists as being mainly of an *a priori* character, even as objection to natural selection, and hence, as offering no positive ground whatever for belief in the inheritance of acquired characters. Of course, at the best, they would only serve to give presumptive support to Lamarckian inheritance.

§ 2. *General Effects and Specific Heredity*

The advocates of the hypothesis of Lamarckian inheritance often fail to distinguish between the effects produced upon the offspring by the general influences of the environment upon the whole organism — *e.g.* malnutrition, toxic agents, such as alcohol, etc. — and the specific modifications of particular parts and functions, arising supposedly from mutilation, use, the stimulation of particular organs, etc. Effects of the former sort are not denied by selectionists; but they claim that this sort of effect produced upon the offspring is rather a disproof than a proof of the Lamarckian view. For example, the effect of alcoholic excess is not an increased tendency in the children to drink alcoholic beverages, — whatever alcoholic tendency there may be in the children is accounted for as already congenital to the parents, — but certain general deteriorating or degenerative changes in the nervous system or constitution of the offspring, manifesting itself in hysteria, scurvy, idiocy, malformations, etc., which the parents did not have at all. Furthermore, the mechanism required to accomplish the two sorts of effect respectively are widely different. The general effects of the first sort, upon the offspring, are due simply to the influences which work upon the organism as a whole, and reach the reproductive cells as well as the body tissues. But to accomplish the transmission of specific modifications of particular parts, a very complex special mechanism would be necessary, whereby the part affected in the parent would impart some sort of special modification to the germ-cells, which would in turn cause the same modification of the same part in the offspring (cf. the address of Sedgwick before the British Association, in *Nature*, Sept. 21, 1899).

It may also be suggested that such a complex mechanism of transmission would be a highly specialized adaptation, and if such a mechanism be necessary to Lamarckian heredity, it would itself have to be accounted for without such heredity. But the rise of complex adaptations is the point at issue.

§ 3. *The Origin of Heredity*

This question takes on considerable importance in view of recent discussion of the origin of heredity itself, in connection with researches into variation. Heredity means, of course, more or less lack of variation — what is called 'breeding true' to stock — from parent to offspring ; it is the opposite of variability, which is departure from the 'true' or like. It has generally been assumed that heredity, at least in the simple form seen in cell-division, — the so-called daughter-cells being parts of the original mother-cell, — was an original property of living matter, and variation from the true was the phenomenon to account for. Recently, however, the theory has been advanced by Bailey (*Plant Breeding*, 1895, and especially *Survival of the Unlike*, 1896) and Williams (*Geological Biology;*

Science, July 16, 1897; *American Naturalist*, Nov.
1898), and advocated independently by Adam Sedgwick (*Nature*, Sept. 21, 1899),[1] that variation is normal,
and that heredity is acquired through the operation of
natural selection restricting and limiting variation to
the extent seen in the relative amount of 'breeding true'
that is actually found in this species or that. It would,
indeed, seem *a priori* more reasonable to ask why such an
unstable compound as protoplasm, acted upon by a complex environment, should not vary (*i.e.*, why it should have
heredity) than the reverse. And, moreover, the compli-
cated apparatus necessary for sexual reproduction and
transmission, itself showing the wide variations it does
in different organisms and in different life conditions,
must, in any case, have been acquired, even though it be
the direct descendant of the earliest forms of cellular
multiplication. Now all of this class of functions — to
come back to our text — emphasizes the requirement of a
theory of the evolution of such a complex apparatus as
that of sexual reproduction and heredity, which does not
assume Lamarckian inheritance — in this case, we may
add, one which does not *assume heredity in order to explain it*.

Again, it has been argued by Weismann and by the
present writer that, if the Lamarckian principle were in
general operation, we should expect to find many functions
which are regularly acquired by each succeeding generation,
such as speech in man, reduced to the stereotyped form of
reflexes or animal instincts.

---

[1] Defrance (*Année Biologique*, V., 1891, p. 375) points out that such a view
was held by Naudin, and refers also to the theory of the origin of heredity held
by Hurst (*Natural Science*, 1890, p. 578); cf. Delage, *Protoplasma*, p. 350.

§ 4. *Lamarckism and Teleology*

The philosophical defence of the Lamarckian principle is usually made from the point of view of teleology, that is, that of a determinate movement in evolution, which is, in some form, the realization of a purpose or end. It is thought that through the accommodations secured by individual animals—provided they be inherited—a determinate direction of evolution toward such a realization is secured; while, on the other hand, the principle of natural selection, working upon 'fortuitous' variations, is called 'blind' and mechanical (cf. the discussion of Ward, *Naturalism and Agnosticism*, Vol. I. Chap. 10).

There seem to the present writer to be certain confusions lurking in such a view. In the first place, it confuses teleology in the process of evolution with purpose in the individual mind. There are two errors here: (1) it is not seen that the evolution process might realize an end or ideal without aid from the individual's efforts or conscious purposes. Indeed, even on the Lamarckian principle, most of the inherited modifications would not be directly due to the individual's purpose or conscious effort, but to semi-mechanical and organic accommodations, and the purpose of the whole could be only partially interpreted in terms of the teleological processes of the individual mind. But those who maintain a general teleological view in cosmology must hold that the cosmic evolution as a whole, and not merely the genesis of certain functions consciously and purposively acquired by the individual, is in some sense purposive. (2) It is not seen that the reverse is also true, *i.e.*, that in spite of purpose in the individual mind, together with the inheritance of acquired modifications in case it be real, the outcome of the evolution movement might still, on the whole, be the same as if it were due to the natural selection of favourable variations from a great many cases distributed fortuitously or by the law of probability. This has been shown, in fact, to be the case in recent investigations in moral statistics ; *e.g.*, suicides are distributed in accordance with the law of probability, and vary with climate, food-supply, etc., in a way which can be plotted in a curve, despite the fact that each suicide chooses to kill himself. That is, the result is as regular and as liable to exact prediction, if we take a large population, as are deaths from disease or accident, or other 'natural' events in which purpose and choice have no part. In such cases, indeed, we have results which are subject to laws as definite as those of mechanics, although the individual data are teleological in the sense of following individual purpose. This case and the reverse, indicated above, show the fallacy of claiming that the exercise of individual purpose is necessarily bound up with a teleological movement in evolution.

§ 5. *Natural Selection not Unteleological*

But there is another supposition open to objection in the view which requires Lamarckian heredity, in order to secure teleology in evolution ; the position that natural selection, working on so-called 'fortuitous' or 'chance' variations, is 'blind' and unteleological. It has been found that biological phenomena — variations in particular — follow the definite law of probability ; in short, that there is no such thing in nature as the really fortuitous or unpredictable. Natural selection, therefore, works upon variations which are themselves subject to law. If this be true, then natural selection may be the method of realizing a cosmic design, if such exists, the law of variation guaranteeing the presence of a fixed proportional number of individuals which are 'fit' with reference to a preëstablished end. All natural processes are subject to law. Design must work out its results by means of natural laws. Why may not the law of probable distribution be the vehicle of such design. Combining this with the result mentioned above, that even moral processes — thus including events in which individual purpose plays a part — are found to be subject to law when taken in large numbers, we are led to the conclusion that the law of probabilities, upon which natural selection rests, is an entirely adequate vehicle of a process of teleology in evolution.

A good illustration may be seen in the use made of vital statistics in life insurance. We pay a premium rate based on the calculation of the probability of life, and thus by observing this law realize the teleological purpose of providing for our children; and we do it more effectively, though indirectly, than if we carried our money in bags around our necks, and gradually added our savings to it. Furthermore, the insurance company is a great teleological agency, both for us and for itself; for it also secures dividends for its stockholders on the basis of charges adjusted to the 'chances' of life, drawn from the mortality tables. Why is it not a reasonable view that cosmic Purpose — if we may call it so — works by similar, but more adequate, knowledge of the whole and so secures its results — *whether in conformity to or in contravention of our individual striving?* Could results so reached be called blind or unteleological? As this point has been put in a recent popular work (*Story of the Mind*, Preface),

" every great law that is added to our store adds also to our conviction that the universe is run through with Mind. Even so-called chance, which used to be the 'bogie' behind natural selection, has now been found to illustrate — in the law of probabilities — the absence of chance. As Professor Pearson has said, 'we recognize that our conception of chance is now utterly different from that of yore . . . what we are to understand by a chance distribution is one in accordance with law, and one the nature of which can, for all practical purposes, be closely predicted.' If the universe be pregnant with purpose, as we all wish to believe, why should not this purpose work itself out by an evolution process under law? — and if under law, why not the law of probabilities? We who have our lives insured provide for our children through our knowledge and use of this law; and our plans for their welfare, in most of the affairs of life, are based upon the recognition of it. Who will deny to the Great Purpose a similar resource in producing the universe and in providing for us all?"

§ 6. *Cosmic Purpose and Law*

Indeed we may go further, and say that this working out of cosmic purpose through some law of the whole, rather than through the individual, is necessary to an adequate theory of teleology as such. In biology the law of 'regression' provides just such a 'governor' or regulator of the process. According to it, individuals which depart widely from the mean do not have proportionate influence on posterity; but there is a regression toward a value which represents the mean attainment of the species up to date. This value is kept fairly constant or gradually advanced. Thus evolution is kept consistently to a determinate direction, and not violently wrenched by what might be called cosmic caprice. It is done by reducing and controlling the influence of individual variations. So it is necessary that the 'choice,' the capricious will or purpose of the individual, should be neutralized if a consistent plan of the whole is to be carried out. Otherwise, it would reflect the irregular variations of our private purposes. This principle of 'regression' or 'conservation of type' holds whether the inheritance of acquired modifications be true or not, — whether the effects of personal effort and purpose be transmitted or not, — and as it deals with all the cases, variations and modifications alike, the purposeful deed of the individual can, in any case, be a factor of but minor importance in the result. Its real importance would depend upon its relation to the whole group of agencies entering into heredity. In so far as individual purpose should be in a direction widely divergent from that of the movement in general, it would, by the law of regression, be largely ineffectual; in so far as it should be in harmony with it, it would be unnecessary and unimportant; although in the latter case, perhaps, taken with the Lamarckian factor, if that be real, it would accelerate biological evolution.

§ 7. *The Place of Individual Purpose in Evolution*

If, after stating the foregoing points as to the relation of the individual's purposes to a possible teleological construction of the evolution movement as a whole, we go on to inquire as to how far the individual may as a fact contribute to the direction of the movement, we recall that the foregoing pages of this work tend to magnify that influence, but to do so under two limitations. The directive influence of the individual's purposes are important either (1) in so far as the accommodations of the individual are *common to a relatively large number*, and so affect the mean values, — which means, really, so far as they are not individual, but for statistical treatment, collective, — or (2) in so far as they affect the progress of the species by *modes of transmission other than those of physical heredity*. The factor called organic selection works, as has been fully shown, through individual modifications; but its rôle is increased by the increase in extent of the accommodations in a group, or by the reduction of the size of the group. A few individuals' accommodations could give a direct turn to the line of progress only in emergencies in which large numbers of those individuals which did not accomplish the accommodation were destroyed. Wherever, however, we find consciousness entering as the vehicle of accommodation, — and this would be the condition of the operation of any factor which could be called by the word 'purpose,' — we find that common widespread forms of accommodation spring up, and the rôle of individual effort, struggle, etc., becomes more prominent as a directive factor.

It is in this latter case, also, that of conscious, somewhat intelligent accommodation, that the second condition mentioned just above comes into play; we find with consciousness the springing up of social modes of transmission: imitation, paternal instruction, and all the processes which we have been calling tradition, social heredity, and transmission. The species may profit by the effects of a single individual's achievements, through social propagation from one individual to another, and through the adoption of social and gregarious modes of behaviour;

and the line of tradition may directly and most strikingly reflect the purposes and attainments of individuals. This has already been touched upon in the section on 'Intelligent Direction' (Chap. X. § 4). Of course so far as it is true that the line of tradition really precedes and sets the direction of the line of physical evolution, by putting a premium upon the educability and plasticity of individuals, in so far the organization of the intelligent purposes of individuals in social and traditional forms would be real and on this definition teleological. But the distinction should be clearly made that this is a factor operative inside the movement itself, through conscious function considered as a character preserved and developed in connection with the brain, and that the larger question of a teleological movement as evolution in general remains still to answer. That such a movement is possible even without this factor is argued above; yet it is natural to look upon the class of phenomena which show the mind taking part in the determination of natural evolution as being in some way in harmony with, or as furthering, the operation of the larger Purpose which a theory of cosmic teleology postulates.

CHAPTER XVII

SELECTIVE THINKING [1]

IN a recent publication [2] I have used the phrase 'selective thinking' in a certain broad sense, and at the same time arrived at a view of the mechanism of the process which seems in a measure in line with the requirements both of psychology and of biology. By 'selective thinking' I understand *the determination of the stream of thought*, considered as having a trend or direction of movement, both in the individual's mental history and also in the development of mind and knowledge in the world. The considerations suggested in the work mentioned are necessarily very schematic and undeveloped, and I wish in this address to carry them out somewhat further.

Looking at the question from a point of view analogous to that of the biologists, when they consider the problem of 'determination' in organic evolution, we are led to the following rough but serviceable division of the topics involved — a division which my discussion will follow; namely, 1. The material of selective thinking (the supply of 'thought-variations' [3]); 2. the function of selection

[1] President's Address, American Psychological Association, Cornell Meeting, December, 1897 (from *The Psychological Review*, January, 1898). The paper aims to present rather a point of view, and to indicate some of the outstanding requirements of a theory, than to defend any hard and fast conclusions.

[2] *Social and Ethical Interpretations*, 1897 (3d ed., 1902).

[3] Wherever the word 'variation' occurs in this chapter, the full term 'thought-variation' should be understood ; this is necessary in order to avoid confusion with the congenital 'variations' of biology.

(how certain variations are singled out for survival); 3. the criteria of selection (what variations are singled out for survival); 4. certain resulting interpretations.

§ 1. *The Material of Selective Thinking*

I suppose that every one will admit that the growth of the mind depends upon the constant reception of new materials — materials which do not repeat former experiences simply, but constitute in some sense 'variations' upon them. This is so uniform an assumption and so constant a fact that it is not necessary to enlarge upon it, at least so far as the growth of our empirical systems of knowledge is concerned. But besides the constantly enlarging and varying actual experiences of the world of persons and things, we have in the imaging functions, taken as a whole, a theatre in which seeming novelties of various sorts are constantly disporting themselves. Seeing further that it is the function of memory, strictly defined, to be true to the past, to have for its ideal the reproduction of experience without variation, it would seem to be to the more capricious exercise of the imaging function which usually goes by the term 'imagination' that we are to look for those variations in our thought contents which are not immediately forced upon us by the concrete events of the real world.

A closer approach may be made, however, to the actual sources of supply of variations in our thought contents, by taking a bird's-eye view of the progress of thought looked at retrospectively; somewhat as the paleontologist puts his fossils in rows and so discovers the more or less consistent trend shown by this line of evolution or by that.

When we come to do this, we find, indeed, certain consistent lines taken by our thought systems in their forward movement — lines which characterize certain more prominent series of stages in the descent or development of the mental life. First, we find the line of knowledges which reveal necessary fact, as we may call it — the line of correspondence between internal relations and external relations upon which Mr. Spencer enlarges, and to which the life of perception and memory must conform. Here there seems to be the minimum of personal selection, because all the data stand on approximatively the same footing, and the progress of knowledge consists mainly in the recognition of reality as it is. Then, second, there is the line of development which shows the sort of concatenation of its members which goes in formal logic by the term 'consistency' and results in some organization. This is often described as the sphere of 'truth' and belief, and is in so far contrasted with that of immediate fact. Third, there is the line of development whose terms show what has been and may be called 'fitness' — a certain very peculiar and progressive series of selections which go to build up the so-called 'ideals,' as in æsthetic and ethical experience.

In addition to these more or less selectively 'determined' lines of orderly arranged materials, there are besides manifold scattered products in the mind at all its levels; and these become especially noticeable when we cast an eye upon the outcome of imagination. We have in so-called 'passive imagination' or 'fancy,' in dreams, in revery, in our air-castle building, untold variations, combinations, and recombinations. The question which comes up for answer in this first survey of these things is this: do the variations by which the lines of consistent, or determined, development are furthered and enriched occur as accidental but happy hits in the overproduced *disjecta membra* of the imaging processes?

I put the question at once in this way in order to come to close quarters with a current way of looking at selective thinking — indeed, about the only current way. To be sure, this question of selection has not been much discussed; but those who have concerned themselves with it have generally been content to say that in imagination, broadly understood, we have the platform on which the true, the good, the valuable thought-variations occur, and from the multitudinous overplus of whose output they are selected.[1]

§ 2. *The Origin of Thought-variations*

This, however, as it seems to me, is quite mistaken. We do not find ourselves acquiring knowledge in our dreams, thinking true in our revery, building up our æsthetic and ethical ideals through castle-building. We do not scatter our thoughts as widely as possible in order to increase the chances of getting a true one; on the contrary, we call the man who produces the most thought-variations a 'scatter-brain,' and expect nothing inventive from him. We do not look to the chance book, to the babbling conversation of society, or to the vagaries of our own less strenuous moods for the influence which — to readapt the words of Dr.

---

[1] This seems to be the assumption, for example, of James (*Principles of Psychology*, II., Chap. XXVIII.). So also Dr. G. Simmel in an article (*Arch. f. sys. Philos.*, I., pp. 34 ff.) which has come to my notice just as this paper goes to print; at least no suggestion appears in his article of any selection except that by movement, to which all thought-variations are alike brought, through what he calls their 'dynamic aspect.' The general positions of Simmel on the origin and meaning of 'truth' are in considerable accord with certain of the conclusions of this address.

R

Stout — 'gives to one of our apperceptive systems a new determination.' On the contrary, we succeed in thinking well by thinking hard; we get the valuable thought-variations by concentrating attention upon the body of related knowledge which we already have; we discover new relations among the data of experience by running over and over the links and couplings of the apperceptive systems with which our minds are already filled; and our best preparation for effective progress in this line or in that comes by occupying our minds with all the riches of the world's information just upon the specific topics of our interest.

All this would lead us to a negative position first — a position which discards the view that the material of selective thinking is found among the richly varied but chaotic and indeterminate creatures of the imaging faculty. Yet it would leave the positive answer to the question of the source of fruitful thought-variations still unanswered.

There are two alternatives still open after the view just mentioned has been discarded; one holding that it is the function of the mind to do its own determining, to think its own apt thoughts, to discover the relations which are true, to bring to the manifold of sense and imagination its own forms, schemata, arrangements of parts, and so to construct its systems of knowledge by the rules of its own inventive power. This theory, it is plain, is analogous to the theory of vitalism, with a self-directing impulse, in biology; and it comes up also rather as an answer to the question as to the forms and categories of mental determination than to that as to the material. For even though the mind has its 'synthetic judgments *a priori*,' as we may say in the phraseology of Kantian philosophy, still the question arises both as to the sources and as to the criteria — the local, temporal, and logical signs — of the empirical data which are utilized in the forms of knowledge. I do not know that any one would be disposed to say that our knowledge of the external world, of the characters of persons, of the truths of history and natural science, are not attained through experience bit by bit ; and the question to which the *a priori* theory gives no answer is : How are these bits found out ? Even given the 'categories,' what sorts of experiences fit the categories, and how is the fitting done ?

§ 3. *The Systematic Determination of Thought*

Leaving for a later section, therefore, the question of the origin of the categories, and reverting to the only remaining real alternative, the first thing to be said is that two limitations confine us in finding the source of the variations which are available for the determination of our thinking, whatever the sphere or line of progress be. First, the new thought-variations, to be candidates for selection, are not mere stray products of fancy; yet second, they are still not outside the problem of selection from variations which arise somehow in the experience of the individual thinker. Having these two limitations full in mind, we find the third alternative — which in my own opinion all the facts go to support — to be this : *the thought-variations by the supply of which selective thinking proceeds occur in the processes at the level of organization which the system in question has already reached — a level which is thus the platform for further determinations in the same system.*

Having stated this general position, we might examine each of the lines or spheres of selective thinking already pointed out; but that does not seem to be necessary. It is just the evident difference between the child and the man, say, that the former proceeds to test data which the latter never thinks of testing. The child thinks the moon may be made of green cheese, that birds may grow on the limbs of trees, that the sun does set around the corner of the world, that eating bread-crusts does make the hair curly; such conceits the man smiles at. The difference is that at the child's level of what we go on to call 'systematic determination,' these are variations of possible value; he has yet to test them; but to the man they are not on the level or platform which his selective thinking has reached; they are not in any sense candidates for selection; they do not even enter into the complexly distributed series of thought-variations within the limits of which his criteria of value and truth lie. Various reasons have been given for this in the literature, and however they differ as explaining principles they are yet severally available as against the theory that all our imaginings afford a chance — and the more, the better the chance — of profit. The untruth of this position is what concerns us.

In getting his information about nature, the child learns by experimenting, as also do the animals. But having learned this or that, *he proceeds on this basis to learn more.* In judging a statement he scouts *in advance* what his lessons have already discredited. In admiring the æsthetic and in adhering to the good, he hesitates only where his sense of worth does not positively go out; what is to him ugly and bad he repudiates with emphasis.

We might take up the parable on the side of brain processes and ask what brain variations give good, true, fit conscious states; and the same would be seen. Suppose, for example, that sane intelligent thinking over the data of the knowledge which one has already acquired involves some sort of coördination of the sensory and motor areas. This coördination is a matter of growth by integration. Variations to be fruitful — whatever be the tests of survival — must be variations in the functioning of this system. Suppose the visual centre rebel and lose its coördination with the motor, or suppose the hearing centre fail of its blood supply, and so drop away from the system; such changes would be gross accidents, temporary inhibitions or diseases, not variations to be selected for the upbuilding and enriching of the system. To be this, brain changes would have to take place in the delicately adjusted processes which constitute the essential coördination in question. I take this case, because, as will appear later, it suggests what is to my mind the real mechanism of selective thinking — coördination of data in the attention, a motor function.

§ 4. *The 'Platform' of Determination*

So far it has seemed that in each case thought-variations must be all at a certain level if any of them are to be available for selection at that level. We may go a step further in the way of defining what is meant by 'level,' or 'platform' of systematic determination.

It is just of the nature of knowledge to be an organization, a structure, a system. There is no such thing as mere 'acquaintance with' anything; there is always — to abuse James' antithesis! — more or less 'knowledge-about.' And the growth of thought is the enlargement of the 'knowledge-about' by the union of partial with partial 'knowledges-about' in a constantly wider and fuller system

of thoughts. Selective thinking is the gradual enlargement of the system, a heaping-up of the structure. If
this be true, a little reflection convinces us that variations
in the items of material merely, in the stones of the structure, in the brute experiences of sense or memory, cannot
be fruitful or the reverse for the system. It is variations
only *in the organization* which can be that. It is the readjustments, the modifications or variations in the 'knowledge-about,' which constitute the gain or loss to thought.
A thousand flashing colours may pass before my eyes, a
thousand brute sounds make a din in my ears, a thousand
personal situations flit through my imagination, a thousand
reports reach me through the 'yellow journals' of the condition of Cuba; but having no tendency or force to work
changes in my organized systems of knowledge, they are
not even possible candidates for my selection. The rich
data of the world and of history might shower upon us;
the music of the spheres might tickle our ears; the ideals
of the Almighty might be displayed before us in colour,
form, and action; but, be we incapable of organizing them,
they are 'as sounding brass and a tinkling cymbal.' The
things of time and eternity may vary infinitely in their
appeals to us, but unless *we vary to meet them* they cannot
become ours. So do we find actually fruitless and barren,
not only the kaleidoscopic changes, the variations on variations, of our dreams and our fancy, but equally so the
pages of mathematical symbols in which we have not been
trained, though they embody the highest thoughts of some
great genius. They do not fit into the coördinations of
knowledge which are ours, nor bring about readjustments
in the arrangements of them. The items, to appeal to me,
must never quite break with the past of my knowledge:

each must have its hand linked with that of the thought which begot it; it must have a 'fringe' if it is to get a lodgement upon the strings of my intellectual loom and stand a chance of being woven into the texture of the carpet which is to cover the upper floor of my mental residence. The burden of mental progress, then, seems to me to lie on the side of the organizing function.

We may believe, therefore, so far as we have gone, that the material available for selective thinking is only of the sort which reflects rearrangements, new adjustments — in short, new 'determinations' — in our organized systems of knowledge; and further that each of such candidates for selection is born, so to speak, at the top of the cone, at the highest floor or level, of its own peculiar system. Other fragments of thought, *disjecta membra* of imagination, lie scattered about the bottom, unavailable and useless. With so much said about the material, we may now go on to consider the function or *process* of selective thinking.

§ 5. *The Function or Process of Mental Selection: the External World*

In the consideration of this problem — of course, the most important one — the advantages of employing the genetic method will become apparent; and it may be well to distinguish the different spheres of mental determination somewhat in the order of their original genetic appearance, the first sphere being that of our knowledge of the external world.

1. The function here is evidently one of an organization of the data of sensation in a way which shall reflect, for our practical purposes, the actual state of things existing in

the world. The selective process must be one which in some way concerns the active life, for it is only through the life of active muscular exertion that the appropriateness of revival processes can be tested. We have here again two alternative views which have been treated in detail in the work, *Mental Development in the Child and the Race;* the one theory, called the 'Spencer-Bain theory,' teaching that all movements showing variation stand on the same footing, and that it is a matter of happy accident as to which of these turns out to be adaptive. Such movements so found out are pleasurable; others, giving pain, are anti-adaptational. Through the mechanism of repetition on the one hand, and of inhibition on the other hand, the former are selected and so survive, and with them survive the feelings, thoughts, etc., which they accompany or secure. The other alternative — advocated in the work mentioned — holds that there is a difference in movements from the start, due to the conditions of waxing and waning vitality from which they spring; pleasure and pain attach respectively to these vital effects of stimulations, and so there is, in each case of a selection of movements, a platform or level of earlier vital adaptations from which the new variations are brought to their issue.[1] This latter theory would seem in so far to get support from the fact brought out above, that such a platform of acquired adaptation — a level of ' systematic determination ' — is present in all selective thinking. This view holds also that such adaptive movements it is which, by their *synergy* or union, give unity and organization to the mental life.

Apart from this, however, the two theories agree in making the selection a matter of motor accommoda-

---

[1] Cf. the expositions in Chap. VIII. § 6, and Chap. IX. § 2.

tion.[1] The system of truths about the world *is a system which it will do to act upon*, both when we take it as a whole and when we go into its details.

Another thing follows, however, — and follows more naturally from the second of the two theories mentioned than from the first, — *i.e.*, that novelty, variety, detail of experience, can be organized in the mental life only in so far as it can be accommodated to by action; if this cannot take place it must remain a brute and unmeaning shock, however oft repeated the experience of it may be. It itself, considered as a thought-variation, as well as the variations in it, would be as if non-existent — altogether without significance for the individual's growth in knowledge. The seat of productive variations, of variations, that is, from which selections are possible, must be on the motor side, in the active life.[2] Only thus could 'internal relations' be established which should be true to or should reproduce 'external relations.'

The point of contrast noted above between the two theories has, however, an additional interest in connection with our present topic; the point that on my theory there is a platform of earlier habitual adjustments from which the variations are always projected. For this transfers the first selective function from the environment to the organ-

---

[1] I am not sure, however, whether Professor Bain does not here leave Mr. Spencer behind. The latter nowhere, to my knowledge, discusses selection in the sense of mental determination, but his insistence upon the direct action of the environment on an organism would seem to require him to hold that the stimulations compelled the organism to accommodate in this direction or that, the motor selection simply coming in after the fact of determination.

[2] By 'motor' is meant vaso-motor and glandular as well as muscular experiences; all of these considered as giving a reflex body of organic contents which cluster up upon incoming stimulations from the external world. It is all afferent, kinæsthetic, in its actual mechanism.

ism, requires the new experience to run the gantlet of habitual reactions or habits which organize and unify the system of knowledges, before it can be eligible for further testing by action. For example, a child cannot play the piano, though he might actually go through a series of movements reproducing those of a skilled performer. The multitude of variations, so far from aiding him, is just the source of his confusion. But he can learn little by little, if he practise faithfully from the platform of the movements of the simple scales and finger exercises which he already knows how to perform.

§ 6. *Tests of Truth in the External World*

The first test, therefore, is that of assimilation to established habits. If we grant this, and also grant that subsequently to this there is a further selection, from such variations, of those which work in the environment, we get a double function of selection : *first, the sort of intra-organic selection called above 'systematic determination,' which is a testing of the general character of a new experience as calling out the acquired motor habits of the organism ;* [1] *and second, an extra-organic or environmental selection, which is a testing of the special concrete character of the experience, as fitted, through the motor variations to which it gives rise, to bring about a new determination in the system in which it goes.*

These selective tests we may call respectively the test of 'habit' and the test of 'accommodation to fact' (the

---

[1] The phrase 'intra-organic selection' suggests (intentionally, indeed, although used here in a purely descriptive sense in antithesis to extra-organic) the process of adaptation called 'Intra-selection' by Weismann and described earlier by Roux under the phrase 'Struggle of the Parts.'

latter abridged to the 'test of fact'). 'These two functions of selection work together. The tests of habit, the intra-organic tests, represent an organization or systematic determination of the things already guaranteed by the tests of fact; and, on the other hand, things which are not assimilable to the life of habit cannot come to be established as intelligible facts. The great difference between the two tests is that that of habit is less exacting; for after a datum has passed the gantlet of habit — or several alternative data have together passed it — it must still compete for survival in the domain of fact.

What, then, do we finally mean by *truth* in the sphere of external knowledge? This, I think: a truth in nature is just something selected by the test of fact (after having passed the gantlet of habit, of course), and then so passed back into the domain of habit that it forms part of that organization which shows the 'systematic determination' of the thinker. What the word 'truth' adds to the word 'fact' is only that a truth is a presentative datum of the intra-organic system which has stood the test of fact *and can stand it again*. A truth is an item of content which is expected, when issuing in movement, to 'work' under the exactions of fact. We speak of a correspondence between the idea and the fact as constituting truth; and so it does. But we should see that a truth is not selected because it is true; *it is true because it has been selected*, and that in both of two ways: first, by fulfilling habit, and second, by revealing fact. There is no question of truth until both these selective functions have been operative. This is to say, from the point of view of motor development, that accommodation always takes place from a platform of habit, and that in the case of the external world our first-hand knowledges arise as reflexes of such accommodations.[1]

§ 7. *Selection of Ideas by Attention*

2. In the life of general and ideal thinking the same questions come upon us. Here we have, it is true, a certain restating of the problem, but it seems that in its essential features the principles already worked out have application. First, as to the platform — for as we saw above, thought-variations to be selected must be projected from a platform of earlier progressive thinking or systematic determination. The platform on the side of function — that is, apart from the content organized — is, I think, *the attention.* The attention is a function of organization, a function which grows with the growth of knowledge, reflects the state of knowledge, holds in its own integrity the system of data already organized in experience. I shall not dwell long upon this, seeing that it will be gen-

---

[1] In *Social and Ethical Interpretations,* Sect. 57, these two phases are generalized as follows: 'With the formula: *what we do is a function of what we think,* we have this other: *what we shall think is a function of what we have done.*' In general conception this is Simmel's position. In the following sentence (of which the passage in the text might almost be considered an English rendering) he is accounting for the 'Harmonie' between thought and action; he says: "Dies (Harmonie) wird erst dann begreiflich wenn die Nützlichkeit des Handelns als der primäre Faktor erscheint, der gewisse Handlungsweisen und mit ihnen die psychologischen Grundlagen ihrer züchtet, welche Grundlagen eben dann in theoretischen Hinsicht als das 'wahre' Erkennen gelten; so das ursprunglich das Erkennen nicht zuerst wahr und dann nützlich, sondern erst nützlich und dann wahr genannt wird" (*loc. cit.,* p. 43). Simmel makes the further argument that in animals of lower orders having senses different or differently developed from ours, the motor accommodations by which the sense organs have arisen must be to different forces and conditions in the environment. So what would be counted 'truth' in the mental systems of such creatures would vary among them and also from our 'truth' (*loc. cit.,* p. 41). An important point of difference between Simmel's view and the writer's is noted below.

erally admitted, I think, that attention is in some way the organizing function of knowledge, and also because further definition — which, moreover, I have attempted elsewhere[1] — is not necessary to our present purpose.

The first selection which thought-variations have to undergo, therefore, if eligibility from this platform be the first condition of final adoption, is in their getting a place in the organization which present attention conditions represent and exact. This is just the condition of things we saw above when we pointed out that it is only the strenuous, hard, and attentive concentration of mind that brings results for the life of thought. Attention is relatively easy, when we let it roam over our old stock in trade; but even then the contrast is striking between the items of knowledge which are held in the system thus easily run through with frequent repetition, and on the other hand those vestigial fragments of representation which do not engage the attention in any system of exercises, and so have no settled place or orderly sequence in our mental life. The latter are not *on the platform ;* the former are. There is always such preliminary 'intra-organic' selection — a set of ready interests, preferences, familiarities, set to catch our new experiences or to reject them. It proceeds by motor synergy or assimilation. Thoughts which get so far in are then candidates for the other selection which the full term 'selective thinking' includes. In order to be really the thought-variations which selective thinking requires, all new items must, in the first place, secure and hold the attention ; which means that they must already enter,

---

[1] *Mental Development in the Child and the Race,* Chap. XV., where it is held that the attention, organically considered, is a habitual motor reaction upon mental contents.

however vaguely, into the complex of earlier knowledge, in order that the habitual motor reflex, which attention is, may be exercised upon them.

In considering in the book cited the empirical complex mental contents which constitute attention,[1] I found it necessary to distinguish three sets of motor events; and I threw them into a certain 'attention formula,' as follows: $Att.$ (attention) $= A + a + a$; the A representing the gross and relatively constant reflex elements which give attention its main sensational character; the a representing the special elements which vary with different classes of experiences, as for example with the different sense-qualities; and the a representing the refined variations which attention to particular objects as such brings out. It is a part of the general analysis of attention which issues in this formula that the state of mind called 'recognition' varies as some or other of these elements of attention are present without variation through repeated experiences. All are present without variation when we recognize a particular object as familiar; there is variation in the a elements only when we are able to place a new object in a familiar class but yet do not find ourselves familiar with it for itself; there are variations in both the a and the a elements when a novel experience simply meets the general requirements of our grosser life of habit, but yet has no place in the organization of our knowledge.[2]

---

[1] *Mental Development in the Child and the Race,* Chap. X. § 3, and Chap. XI. § 2.

[2] Thus the animal instincts show gross motor reactions upon the objects which call them out, and it may be that the only differentiation of the objects possible to the creature is just that supplied by his differentiated instinctive attitudes including the attention.

This analysis enables us to see more clearly the meaning of the 'platform' from which thought-variations must be projected to be real candidates for selection in the life of attention. The experience which does not even bring out the constant A elements is merely a brute shock and not 'knowledge-about,' seeing that in these elements, which are necessary to all attention, we have necessary motor adjustments in which accommodation to the external world consists. Such 'shocks' do not reach the platform.

Further, those experiences which do involve the A elements must also, at least in selective thinking, have some sort of a element and a elements with them; seeing that, in the realm of thought, attention which is not concrete involves no specific determination.[1] The study of the child shows that the differentiation of the a from the a elements is a gradual thing, the first knowledge being of a 'vaguely universal' sort (an expression of Royce's; the same thing has been called by the present writer 'the general of the first degree'). Psychologically, therefore, the platform upon which the new knowledge is to be secured is that of a sense of familiarity toward an experience, at least in the unrefined way which the child's 'vague universal' illustrates. The apprehension of a new truth is always either the consciousness of an identity, in which case it is treated as an old truth in all respects, or it is in some measure subsumed under an old truth, when it illustrates class recognition. And it follows as to our platform that any new knowledge, to be selected and held as such, must be

---

[1] We may note, however, the familiar fact that the concrete content on which attention is fixed is often only a *point d'appui*, a symbol, verbal or other, which, on the organic side, merely opens a channel for the discharge of the larger whole of attitudes (the a elements) which general and class notions presuppose.

capable of the sort of subsumption which class recognition is. This gives, in the sphere, of general thought, the analogue of the assimilation to habit which we found necessary to the establishing of the platform of progressive determination in the case of knowledge of external objects. *The two cases taken together, therefore, constitute the function of 'systematic determination.'*

§ 8. *Variations in Attention and the Environment of Thought*

But this is not yet selective thinking. The selection of the particular concrete datum is more; it is *an affair of the selection of variations in the attention complex*, after the datum has passed muster in the systematic determination. It is an affair of the variations of the *a* sort, at the crest, so to speak, of the attention movement. How, then, are these selected?

It is, I think, a process analogous to that which holds for muscular accommodation by adjustment to the environment, *i.e.*, it is a case of 'functional selection from overproduced movements.' It is here, as there, the environment's turn to get in its work, after the organism has had its turn. Yet here, as there, we must be careful to have a clear understanding of what the environment is.

The environment is here *the whole of knowledge not possessed by the individual thinker;* that is, the whole of the social store of opinions, beliefs, reflections, judgments, criticisms, etc., within which the individual displays his reasonable activities. The selection of thoughts as valid is analogous of the selection of facts as true. Apart from the direct necessity of accommodation and recognition which the physical enforces upon us, and which constitutes the selection of certain facts from all those possible but pseudo-facts which our habitual reactions might allow to pass — apart from such physical facts, all truths are selected by a testing in the social environment, from the many pseudo-truths which have passed the gantlet of our habitual attention reactions.

To illustrate: we see a vague outline in the dusk; it might be a man, a beast, a tree-stump, so far as our present adaptive attitudes and recognitions avail to define it. To decide which it is and so to select one alternative as true, we put it to the physical tests of nearer approach, touch, hearing, etc. Here we have first the platform, then the selection by further action. So in thinking: we hear, let us say, a report that a friend is dead; he may have died by accident, by poison, by fire, so far as our information goes. We find out the truth, however, by getting information from some one who knows. Here, again, is the platform with its alternatives (variations), and then the selection by a social appeal. In the case of a scientific invention the part which can be attested by an appeal to fact is so tested, but the part which still remains hypothetical is so far liable to social confirmation that the inventor expects at least that others will judge as he judges.

The use of the word 'judge' in the last sentence serves to suggest certain further considerations, which show the social appeal in operation, and, at the same time, give evidence that it is this appeal which constitutes the resource in selective thinking in the higher and more ideal spheres. These considerations may be presented under the third heading, dealing with the criteria of 'fitness' of thoughts.

s

§ 9. *What constitutes Fitness in the External World?*

By criteria here is meant not so much objective criteria — marks or characters of this or that experience — as criteria of survival, *i.e.*, the tests or qualifications which new items of experience must fulfil if they are to be given a permanent place in the organization of knowledge. This involves the question of objective criteria, to be sure; but we may be able to find some general qualification under which the special criteria of the different provinces of knowledge may be viewed. Our question may be put in the familiar terms of an analogous biological problem, if we ask: when a particular truth has been shown by selection to be such, *why was it found fit to survive?*

In answer to this question we may say at once, concerning knowledge of the external world, that the motor accommodations by which the selective process proceeds are, by the conditions of the environment, *of necessity made in this direction or that.* The reason a given movement is fit is because it actually reports fact. The dictum of the environment is: accommodate to *xyz* or die in the attempt! The facts are there; nature is what it is; the adjustments are such just because they are fit to report a state of facts. The environment in which the accommodations take place, and to which they constitute adjustments, is the control factor, and its facts constitute the only reason that the selections are what they are. The criterion here, therefore, is simply the adaptive aspect of the movement, as reporting fact. It can be determined in each case only after the event; that is, after the selection has taken place.

But even in this lower sphere, where the exigencies of

the physical environment are the control-factor in the selective process, we find the further result that the preservation of the fact selected depends upon its having already been assimilated to the organized habits of the individual. As knowledge it becomes part of a system; it is added to the platform from which subsequent selections are made; and it thus carries forward the 'systematic determination' of thought. In this way *the organism gradually reproduces in its own platform of determination the very criteria of selection at first enforced only by the environment.* We should expect to find in consciousness some general colouring due to the attitude which the platform of systematic determination requires — an attitude of welcome, of hospitality, of indorsement, in short, of *belief*—toward those facts which have passed through the selective processes, have been added to the organization of knowledge, and have acquired the *cachet* of familiarity.

We need not stop to argue that it is right to apply the term 'belief' to this sense of the internal fitness of experiences after their selection; the implied converse proposition, *i.e.*, that belief is a motor or active attitude, has been ably argued by Bain, James, and others. It is also advocated in the writer's *Feeling and Will.* But whatever we call it, there is the fact — and that is what I wish to emphasize under whatever name — that even in our knowledge of nature the individual *gradually builds up internally* the criteria of selection; and as his experience extends ever more widely afield from the brute resistances, strains, and contacts with things, he becomes a more and more competent judge for himself of the value of variations in his thoughts. Here is what is essential in it all:

the sense of values has grown up all along under the actual limitations of control from the imperative selective conditions of the environment, and if one make use of his criteria of selection beyond the teachings of his experience it is only by means of those general rules which are implicit in the systematic determination itself.[1]

§ 10. *The Fitness of Ideas: the Social Environment*

Turning now to the great platform of attention, we find an analogous state of things, and the analogy really turns out to be identity of process, thus providing a strong argument for the view that the social criterion of selection is here the true one.

In the first place, we have to recognize that *in all thinking whatsoever as such* — even in our thinking about the external world when viewed not as motor accommodation, but as a system of organized truths — *the environment is social.* For we may ask : what does environment mean? Does it not mean that set of conditions which runs continuously through the individual who is said to be in the environment ? The physical environment is such because its conditions are those of motion, while the organism moves. The environment of thought can only be thoughts ; only processes of thought can influence thoughts and be influenced by them. The sources from which spring items in the world of thought are ordinarily centres of thought —

---

[1] Such as the laws of identity (motor habit), consistency (motor assimilation), sufficient reason (accommodation to the item selected), etc.; cf. *Mental Development in the Child and the Race*, Chap. XI. § 1. By these I think it possible to account for the so-called 'analytical processes,' which have to deal with relationships inside the whole of the systematic determination, on which see further below.

minds, either one's own or some one's else. So the environment must be the persons about the thinker. They constitute his environment; they give him conditions to react upon; *they are the control factor in his higher selective thinking*, just as the world of things is the control factor in his life of sense perception. I know that it is through their life of action — mainly indeed, the speech functions — that he realizes their thought, and it is through *his* life of action that he reacts upon their thought and exhibits his; but even in knowledge of the external world of signs, expressions, etc., we have to say that movement must be reduced to some form of thought in order to be organized in our knowledge. And as soon as we get out of the sphere of knowledge of the world of things, and ask how knowledge can proceed without the selective control of physical fact upon movement, we have to say that if selection is to have reference to any environment at all it must have reference to an environment of thinking. Apart from theory, however, the social life is as a matter of fact the environment of our thinking; in the recent book already referred to,[1] there is cited much evidence to show that the child organizes his thoughts with constant reference to the control which the social environment enforces.

So we have found that each group of thought-variations, to be candidates for selection, must be projected from a platform of acquired knowledge, represented on the motor side by certain elements in the attention complex which give the sense of familiarity, class identity, general truth, or vague universality. This is the platform of systematic determination through the attention. Now, why not stop here? Because when a new thing comes, this does not

[1] *Social and Ethical Interpretations.*

suffice to secure those more refined elements of the attention complex which determine a new concrete fact. On the contrary, many alternative determinations, all of them answering the demands of the platform of vague generality, might be forthcoming and the mind might rest in any one of them. Note the child's long-continued and fanciful speculations about the simplest events in the household. What must now be had is just the selective control of an environment in which such variations can be brought to a test; and to the child this is the environment supplied by the persons who know more than he does. To them he normally appeals, almost invariably accepts their decisions, and finds certain of his alternatives thus selected, by what is to him as direct an adjustment to fact as are the selections of his movements by accommodation to that other environment, the world of things. Every new piece of knowledge needs this confirmation just in so far as the systematic determination by which it is brought to the bar of selection leaves the concrete filling of the event indefinite; that is, in so far as various alternatives or variations might be brought into selective rivalry with it.

But then — and this is a vital fact in the growth of the individual — this selection by a social criterion *becomes personal to the learner through his renewed action.* The selected functions, with their knowledge contents, *are added to the organization within, so that the 'systematic determination' of the future is influenced by the assimilation of each new selected element.* Thus the inner attitude which the individual brings to his experience undergoes gradual determination by the continued selective action of the social environment. He himself comes more and more to reflect the social judgment in his own systematic determination of knowledge; and there arises within himself a criterion of a private sort which is in essential harmony with the social demand, because genetically considered it reflects it. The individual becomes a law unto himself, exercises his private judgment, fights his own battles for truth, shows the virtue of independence and the vice of obstinacy. But he has learned to do it by the selective control of his social environment, *and in his judgment he has just a sense of this social outcome.*

In the work referred to I have dwelt at length upon the actual facts of this educative dependence of the individual upon social lessons. The aspect to be emphasized here is the selective aspect, *i.e.*, the truth that the internal criterion is, so far as it goes, always in fact the primary criterion in our thinking; but that in its origin the relation is quite the reverse; and, further, that the individual's judgment is liable all the time to the final selective revision of the social voice. This shows itself most markedly in those ideal states of mind in which the direct control of objective fact is lacking and where the private determination is more or less explicitly accompanied by a sense of 'publicity'—a sense that the public judgment is implicated with one's own in the approval or disapproval of this act or that. In our ethical judgments I think this ingredient is unmistakable.

It remains only to say again that in the state of mind called belief, mental indorsement, and in particular cases judgment, we have the actual outgoing of this systematic determination upon the details of experience. All judgments in experience are, I think, acts of systematic determination, acts of taking up an attitude, of erecting a platform from which new things, to be eligible for selection, must be viewed. The details of organization, thus gradually built up, show the relationships of our theoretical thought; and these relationships are valid since they reveal the motor organization which has accrued to the attention complex. The data of fact or objective truth are the items which have passed through the selective ordeals.[1]

§ 11. *Summary*

The general conclusions which the sketchy development so far made would suggest may be stated in summary form before we go on to note some further points of interpretation in the last remaining section. These conclusions are as follows. Selective thinking is the result of motor accommodation to the physical and social environment; this accommodation taking place in each case, as all motor accommodation does, from a platform of earlier 'systematic determination' or habit. In the sphere of the physical environment as such, the selection is from

---

[1] 'So erzeugen sich für unser Denken, gemäss dem Nützlichkeitsprincip, gewisse Normen seines Verhaltens, durch welche überhaupt erst das zustande kommt was wir Wahrheit nennen, und die sich in abstracter Formulirung als die logischen Gesetze darstellen' (Simmel, *loc. cit.*, p. 45). It is here that the difference between Simmel's view and my own may be noted. He makes (so far as the undeveloped form of his article justifies an interpretation) the function of movement that of giving 'truth' to thought-variations *already present*. The 'dynamic aspect' in its issue secures the selection of the ready-formed 'presentative aspect.' This I hold to be true (when supplemented by the 'systematic determination' of the variations on a platform) of presentative data, wholes, or facts as such. But there still remain the determining effects of the motor selections themselves upon the systematic determination. The synergies, inhibitions, etc., of the new motor accommodations with old habits produce changes in the *organization or relationships of the data* and give rise to theoretical and analytical 'validity' in our knowledge, which differs (as Simmel himself points out, and as Urban has independently suggested) from the objective 'truth' of given data or 'wholes.'

overproduced movements projected out from the platform of the habitual adaptations of the members brought into play ; in the sphere of the social environment it consists in the accommodation of the attention, secured by the over-production of motor variations projected from the platform of the habitual attention complex. The presentations from which the selected motor variations issue are believed or called 'true,' while the organization which the motor complex gradually attains holds the data of ·knowledge in relations of theoretical and analytical 'validity.' In the case of physical selection the internal organization represented by systematic determination gradually serves to free the organism from direct dependence upon the control of the environment ;[1] in the intellectual life this is even more true, the development of the individual's judgment growing more and more independent of social control as progress is made in the 'systematic determination.'

This general sameness in the operation of selection in the two spheres is what we should expect if the method of motor accommodation be what I have described as the imitative or 'circular' reaction. For it is just through reactions of this type, with the antithesis between pleasure and pain by which they are furthered and maintained, that motor accommodations are all the while passed over to the domain of habit, that is, integrated in the system of 'intra-organic determinations.' Thus organized knowledge in all its development may be looked upon as due to the *synergies* of motor processes selected as accommodations to the world of things and persons.[2]

[1] This is seen in psychogenetic evolution in the rise of memory, thought, etc., considered as variations, which constitute a more or less self-subsistent and independent 'mental life.'

[2] Argued in detail in *Mental Development.*

§ 12. *Some Fragmentary Interpretations*

In the way of showing certain general bearings of the position now taken, — bearings which the limits of this address do not enable me to amplify in any detail, — I may go on to indicate the points which follow. They are suggestions toward a broader union of points of view.

1. It will be seen that the position now taken up preserves what may be called the 'utility' criterion of survival through the whole progress of knowledge. The acts of selection are never independent of control from experience, however adequate they may be *within* this control; for the internal or systematic determination, while always the platform of variation, is yet never the final agent of concrete selection. To be sure, the individual's judgment, his sense of reality and truth, becomes more independent or self-legislative, as we have seen; but this, when genetically considered, is both the outcome and the evidence of the control which the environment has all along exercised. Even though we assume certain innate norms of selection which the individual directly applies, still these norms must not only lead to workable systems of knowledge in the world of active experience, but they must also in their origin have been themselves selected from variations, unless, indeed, we go back to a theory of special creation with preëstablished harmony.[1] But if we admit that they are themselves selected variations, then we find no way to account for their selection except that by accommodation to the physical and social environments.[2] We thus preserve

---

[1] Cf. the following chapter for some criticisms of this theory.

[2] Simmel makes the analogous argument (*loc. cit.*, p. 45) that even if we had on *a priori* stock of knowledge, a selection of movements would still have to be made for practical life, and a system of 'truths' would still be built up thereby.

the utility criterion, therefore, even though we may not accept the precise method of selection portrayed above.

2. This does not tend, however, to give support to Mr. Spencer in looking to 'race-experience' for the origin of the categories of knowledge. Spencer's theory has been admirably criticised by Professor James, who thinks that the forms of knowledge must be looked upon as variations, not as accumulations from the repeated impressions of the environment. In support of James' argument we may add — what to me seems an insurmountable objection — that Spencer's position requires the transmission of such impressions by heredity [1] — a notion which James was one of the first to combat and a claim for which no evidence is forthcoming. The position developed above assumes variations, with constant systematic determination in the individual's experience ; in this the control of the environment is reflected. We then need a theory of evolution which will account for the determination of race-progress in the lines thus marked out by the individual.

3. This requirement seems to be met by the theory of Organic Selection, developed in the preceding pages, considered as supplementary to natural selection in the way of securing lines of determinate evolution. According to this view, those individuals which successfully accommodate to the environment live and keep alive, through heredity, the congenital variations which they exhibit. To these are added further congenital variations which are again selected. Thus variations are secured in definite lines in

[1] Cf. the dogmatic utterance of Wundt *apropos* of instinct : "The assumption of the inheritance of acquired dispositions or tendencies is inevitable, if there is to be any continuity of evolution at all. We may be in doubt as to the extent of this inheritance; we cannot question the fact" (*Human and Animal Psychology*, p. 405). Hoc atque anno 1892 !

a series of generations — lines which produce the determinations first secured in the individual under the control of the environment. On this view, there would be a constant selection of individuals by natural selection, from a platform of organic selection which is analogous to the platform of 'systematic determination' in the individual. Race evolution would thus, on the whole, conform to the exigencies of experience, which would seem to be directly transmitted by heredity, while due, in truth, to a series of variations accumulated by natural and organic selection.

4. Furthermore, the content of the intellectual and social environment is kept constant by the handing down of tradition through 'social transmission,' and the same demands are thus made upon the individuals of each new generation, both as to their concrete selections under the control of the environment and as to the forms in which the 'systematic determination' of knowledge is cast.

5. Finally, the 'systematic determination' of the individual thinker reflects, on the subjective side, his *sense of self*. Judgment is the *personal* indorsement of the data of knowledge. Belief is a *personal* attitude. *The person is the whole of the organization of knowledge;* and as the social criteria of selection — and the social data selected — play an essential *rôle* in the process of systematic determination, as explained above, so the person is always social. This I have developed at length elsewhere.[1]

---

[1] In *Social and Ethical Interpretations;* in the third edition, Appendix K, a further note is made on 'Selective Thinking' in reply to a criticism by Dr. B. Bosanquet.

CHAPTER XVIII

The Origin of a 'Thing' and its Nature [1]

THE present growing interest in genetic problems, as well as the current expectation that their discussion may render it necessary that certain great beliefs of our time be overhauled — these things make it important that a clear view should be reached of the sphere of inquiry in which questions of origin may legitimately be asked, and also just what bearing their answer is to have upon the results of the analytic study of philosophy.

We already have, in several recent publications, the inquiry opened under the terms 'origin *vs.* reality' — or, in an expression a little more sharp in its epistemological meaning, 'origin *vs.* validity.' I should prefer, in the kind of inquiry taken up in this paper, to give a wider form to the antithesis marked out, and to say 'origin *vs.* nature,' meaning to ask a series of questions all of which may be brought under the general distinction between the 'how' of the question: how a thing arose or came to be what it is; and the 'what' of the question: what a thing is.

§ 1. *What is a Thing?*

Well, first, as to 'what.' Let us see if any answer to the question 'What is it?' can be reached, adequate to our needs, in any case of genetic inquiry. It seems that the

[1] Paper presented to the Princeton Psychological Seminar in May, 1895, slightly revised (from *The Psychological Review*, November, 1895).

philosophy of to-day is pretty well agreed to start analysis of a thing inside of the behaviour of the thing. A 'thing' is first of all so much observed behaviour. Idealists pass quickly over the behaviour, it is true; it is too concrete, too single, for them; it is not to them a thing, but a 'mere thing.' But yet they do not any longer allow this 'mereness' to offend them to the extent of drawing them off to other fields of exploration altogether. They try to overcome the 'mereness' by making it an incident of a larger fulness; and the 'implications' of the thing, the 'meaning' of it 'in a system'—this 'shows up' the mereness, both in its own insignificance and in its fruitful connection with what is universal.

So we may safely say of the idealist, that if he have a doctrine of a 'thing,' it must, he will himself admit, not be of such a thing that it cannot take on the particular form of behaviour which the one 'mere thing' under examination is showing at the moment. There must, in short, be no contradiction between the 'real thing' and the special instance of it which is found in the 'mere thing.'

He, the idealist, therefore, is first of all a phenomenist in constructing his doctrine of the real; the 'what' must be, when empirically considered, in some way an outburst of behaviour.

Now the idealist is the only man, I think, of whom there is any doubt in the matter of this doctrine of behaviour, except the natural realist, who comes up later. Others hold it as a postulate since Lotze, and later Bradley, did so conclusively show the absurdity of the older uncritical view which held, in some form or another, what we may call the 'lump' theory of reality. A thing cannot be simply a lump. Even in matter—so we are now taught by the physicists — there are no lumps. To make a thing a lump — not to cite other objections to it — would be to make it impossible that we should know it as a thing. So all those doctrines which I have classed as other than idealistic accept, and have an interest in defending, the view that the reality of a thing is presented in its behaviour.

§ 2. *A Thing is Behaviour; the 'What' and the 'How'*

So setting that down as the first answer to the 'what' question, we may profitably expand it a little. The more we know of behaviour of a certain kind, then the more we know of reality, or of the reality, at least, which that kind of behaviour is. And it is evident that we may know more of behaviour in two ways. We may know more of behaviour because we take in more of it at once; this depends on the basis of knowledge we already have — the relative advance of science in description, explanation, etc., upon which our interpretation of the behaviour before us rests. In the behaviour of a bird which flits before him, a child sees only a bright object in motion; that is the 'thing' to him. But when the bird flits before a naturalist, he sees a thing whose behaviour exhausts about all that is known of the natural sciences. Yet in the two cases there is the 'thing,' in just about the same sense.

When we come, moreover, to approach a new thing, we endeavour, in order to know what it is, to find out what it is doing, or what it can do in any artificial circumstances which we may devise. In as far as it does nothing, or as far as we are unable to get it to do anything, just so far we confess ignorance of what it is. We can neither summon to the understanding of it what we have found out about the behaviour of other things, nor can we make a new class of realities or things to put it in. All analysis is but the finding out of the different centres of behaviour which a whole given outburst includes. And the whole, if unanalyzable to any degree, is itself a thing, rather than a collection of things.

But the second aspect of a thing's reality is just as important. Behaviour means in some way change. Our lump would remain a lump, and never become a thing if, to adhere to our phenomenal way of speaking, it did not pass through a series of changes. A thing must have a career; and the length of its career is of immediate interest. We get to know the thing not only by the amount of its behaviour, secured by examining a cross-section, so to speak, but also by the increase in the number of these sections which we are able to secure. The successive stages of behaviour are necessary in order really to see what the behaviour is. This fact underlies the whole series of determinations which ordinarily characterize things, such as cause, change, growth, development, etc., as comes out further below.

The strict adherence to the definition of a thing in terms of behaviour, therefore, would seem to require that we waited for the changes in any case to go through a part at least of their progress; for the career to be unrolled, that is, at least in part. Immediate description gives, as far as it is truly immediate, no science, no real thing with any richness of content; it gives merely the snap-object of the child. And if this is true of science, of everyday knowledge which we live by, how much more of the complete knowledge of things desiderated by philosophy? It would be an interesting task to show that each general feature of the 'what' in nature has arisen upon just such an interpretation of the salient aspects presented in the career of individual things. But this would be to write a large and most difficult chapter of genetic philosophy.

Our second point in regard to the 'what,' therefore, is that any 'what' whatever is in large measure made up of *judgments based upon experiences of the 'how.'* The fundamental concepts of philosophy reflect the categories of origin, both in their application to individuals — to the 'mere thing' — and also in the interpretation which they have a right to claim; for they are our mental ways of dealing with what is 'mere' on one hand and of the final reading of reality which philosophy makes its method. Of course the question may be asked : how far, origin ? — that is, how far back in the career of the thing is it necessary to go to call the halting-place 'origin'? This we may well return to lower down; the point here is that origin is always a reading of part of the very career which is the content of the concept of the nature of the thing.

§ 3.   *The 'What' and the 'How' of Mind*

Coming now closer to particular instances of the 'what,' and selecting the most refractory case that there is in the world, let us ask these questions concerning the *mind.* I select this case because, in the first place, it is the case urgently pressing upon us ; and, second, because it is the case in which there seems to be, if anywhere, a gaping distinction between the 'what' and the 'how.' Modern evolution claims to discuss the 'how' only, not to concern itself with the 'what'; or, again, it claims to solve the

T

'what' entirely by its theory of the 'how.' To these claims what shall we say?

From our preceding remarks it seems evident that the nature of mind is its behaviour generalized; and, further, that this generalization necessarily implicates more or less of the history of mind; that is, more or less of the career which discloses the 'how' of mind. What further can be said of it as a particular instance of reality?

A most striking fact comes up immediately when we begin to consider mental and with it biological reality. The fact of growth, or to put the fact on its widest footing, the fact of *organization*. The changes in the external world which constitute the career of a thing, and so show forth its claim to be considered a thing, fall under some very wide generalizations, such as those of chemistry, mechanics, etc.; and when the examination of the thing's behaviour has secured its description under these principles in a rather exhaustive way, we say the thing is understood. But the things of life, and the series of changes called organic which unroll its career, are not yet so broadly statable. When we come to the mind, again, we find certain very well made out generalizations of its behaviour. But here, as in the case of life, the men who know most have not a shadow of the complacency with which the physicist and the chemist categorize their material. It is for this reason, I think, in part, that the difference between the two cases gets its emphasis, and the antithesis between origin and nature seems so necessary in one case while it is never raised in the other. For who ever heard an adept in natural science say that the resolution of a chemical compound into its elements, thus demonstrating the elements and law of the origin of the 'thing' analyzed, did not solve the question of its nature, as far as science can state a solution of that question?

§ 4. *The 'Prospective' and the 'Retrospective'*

But we cannot say that the whole difference is one of greater modesty on the part of the psychologists. The facts rather account for their modesty. And the prime fact is one formulated in more or less obscurity by many men, beginning with Aristotle: the fact, namely, that organization, considered as itself a category of reality, never reaches universal statement in experience. To confine the case at first to vital phenomena, we may say that to subsume a plant or animal under the category of organization is to make it at once to a degree an x; a form of reality which, by right of this very subsumption, predicts for itself a phase of behaviour as yet unaccomplished — gives a prophecy of more career, as a fact, but gives no prophecy (apart from other information which we may have) of the new phase of career in kind. Every vital organization has part of its career yet to run. If it has no more career yet to run, it is no longer an organization; it is then dead. It then gets its reality exhausted by the predication of the categories of chemistry, mechanics, etc., which construe all careers retrospectively. A factor of all biological and mental realities alike is just this element of what has been elsewhere called 'Prospective Reference.'[1] In biology it is the fact of Accommodation; in psychology it is the fact found in all cases of Selection — most acute in Volition.

And it does not matter how the content in any particular filling up of the category may be construed after it

[1] See *Mental Development*, Chaps. VII., XI.

takes on the form of accomplished fact — after, *i.e.*, it becomes a matter of 'retrospect.' All constructions in terms
of content mean the substitution of the retrospective
categories for those of prospect ; that is, the construction
of an organization after it is dead, or — what amounts to
the same thing — by analogy with other organizations
which have run down, or died, in our experience. Suppose, for example, we take the construction of the category
of accommodation, in each particular instance of it, in
terms of the ordinary biological law of natural selection —
an attempt made by the present writer under the statement of 'functional selection'[1] — and so get a statement of how an organism actually got any one of the
special adjustments of its mature personal life. What,
then, have we done? I think it is evident that we have
simply resorted to the 'retrospective' reference ; we have
changed our category in the attempt to get a concrete filling for a particular case *after it has happened.* To adopt
the view that the category of organization can be in every
case filled up with matter, in this way, does not in any sense
destroy the prospective element in the category of organization ; for the psychological subtlety still remains in mind in
the doing of it, either that the event must be awaited to
determine the outcome, and that I am agreeing with myself and my scientific friends to wait for it, or that we are
solving this case by others for which we did wait. A
good instance of our mental subtleties in such cases is seen
in the category of 'potentiality' considered lower down.
The extreme case of the reduction of the categories of

---

[1] In accordance with which the organism's new accommodations are
selected out of movements excessively produced under pleasure-pain stimulation. *Ibid.*, pp. 174 f. (cf. pp. 45, 94 f. above).

prospective reference to those of retrospect is evidently the formula for probabilities. I do not see how that formula can escape being considered a category of retrospect, applied to material which does not admit of any narrower or more special retrospective formulation.

Now the inference from this is that our predicate 'reality,' in certain cases, is not adequately expressed in terms of the experienced behaviour of so-called real content. The very experience on the basis of which we are wont to predicate reality testifies to its own inadequacy. There is no way to avoid the alternatives that either the notion of reality does not rest upon experiences of behaviour, or that the problematic judgments based upon those experiences of progressive organization which we know currently under the term 'development,' are as fundamental to these kinds of reality as are those more static judgments based on history or origin.

§ 5. *Probability and Design*

It may be well, in view of the importance of this conclusion, to see something more of its bearings in philosophy. The historical theories of 'design,' or teleology in nature, have involved this question. And those familiar with the details of the design arguments pro and con will not need to have brought to mind the confusion which has arisen from the mixing up of the 'prospective' and 'retrospective' points of view. Design, to the mind of many of the older theistic writers, was based upon relative unpredictability — or better, infinite improbability. Such an argument looks forward; it is reasoning in the category of organization, and under the 'prospective' reference. The organization called mental must be appealed to. What, was asked, is the probability of the letters of the *Iliad* falling together so as to read out the *Iliad*? The opponents, on the other hand, have said : Why is not the *Iliad* combination as natural as any other?—one combination has to happen ; what is to prevent this? If a child who cannot read should throw the letters, the *Iliad* combination is no more strange to him than any other. These men are reasoning in the retrospective categories. They are interpreting facts. The fault of the latter position is that it fails to see in reality the element of organization which the whole series when looked at from the point of view of the production of the *Iliad* requires. It is true that the *Iliad* is one of an infinite number of possible combinations ; but it is also true that Homer did not try the other combinations before hitting upon the *Iliad*.

What would really happen, we may say, if the child should throw the *Iliad* combination, would be that nature had produced a second time a combination once before produced (in the mind of Homer, and through him in ours) without fulfilling all the other combinations — an infinite number — which have a right to be fulfilled before the *Iliad* combination be reproduced. And it is the correspondence of the two — apart from the meaning of the *Iliad* at its original production — which would surprise us.

But it is clear that the additional element of organization needed to bring nature into accord with thought and which the postulate of design makes in reaching a Designer — this is not needed from the mere historical or retrospective examination of the facts. In other words, if the opponents of design are right in holding to a complete reduction of organization to retrospective categories, they ought to be able to produce intelligible results by throwing a multiple of twenty-six dice each marked with a letter of the alphabet.

The later arguments for design, therefore, which tend to identify it with organization, and to see in it, so far as it differs from natural law, simply a harking forward to that career of things which is not yet unrolled, but which when completely unrolled will be a part of the final statement of origins in terms of natural law — this general view has the justification of as much criticism as has now been stated.

§ 6. *Design is Genetic*

And, further, it is clear that the two opposed views of adaptation in nature are both genetic views — instead of being, as is sometimes thought, one genetic (that view which interprets the adaptation after it has occurred) and the other analytic or intuitive (that view which seeks a beforehand construction of design). The former of these is usually accredited to the evolution theory; and properly so, seeing that the evolutionist constantly looks backward. But the other view, the design view, is equally genetic. For the category of higher or mental organization by which it proceeds is just as distinctly an outcome of the movement or drift of experience toward an interpretation of career in terms of history. Teleology, then, when brought to its stronghold, is a genetic outcome, and owes what force it has to the very point of view that its most fervent advocates — especially its theological advocates — are in the habit of running down. The consideration of the stream of genetic history itself, no less than the attempt to explain the progress of the world as a whole, its career, leads us to admit that the real need of thinking

of the future in terms of organization is as great as the
need of thinking of the past in terms of natural law.   The
need of so-called mental organization or design is found
in the inadequacy of natural law to explain the further
career of the world, and its past career also as soon as we
go back to any place in the past and ask the same question there.   It would be possible, also, to take up the last
remark for further thought, and to make out a case for the
proposition that the categories of ' retrospective ' thinking
also involve a strain of organization — a proposition which
is equivalent to one which the idealists are forcibly urging
from other grounds and from another point of view.
Lotze's argument to an organization at the bottom of
natural causation has lost nothing of its power.   Viewed as a
category of experience, I am unable to see the force of the
assumption tacitly made by the Positivists, and as tacitly
admitted by their antagonists, that causation is to be ultimately viewed entirely under such retrospective constructions as ' conservation of energy,' etc.   Such constructions
involve an endless retrospective series.   And that is to
say that the problem of origin is finally insoluble.   Well,
so it may be.   But yet one may ask why this emphasis of
the 'retrospective,' which has arisen in experience with
only the basis of experience that the 'prospective' also
has?   It may be a matter of taste ; it may be a matter of
'original sin.'   But if we go on to try to unite our categories of experience in some kind of a broader logical
category, the notion of the Ultimate must, it would seem,
require both of the aspects which our conception of reality
includes : the 'prospective' no less than the 'retrospective.'   Origins must take place continually as truly as
must sufficient reasons.   The only way to avoid this is to

say that reality has neither forward nor backward reference. So say the idealists in getting thought which is not in time. But be that as it may, we are dealing with experience, though for myself, I must say, thought which looks neither backward nor forward is no thought at all.

§ 7. *The Natural History of the Categories*

Another subtlety might raise its head in the inquiry whether in their origin all the categories did not have their 'natural history.' If so, it might be said, we are bound, in the very fact of thinking at all, to give exclusive recognition to the historical aspect of reality. But here is just the question: does the outcome of career to date give exhaustive statement of the idea of the career as a whole? There would seem to be two objections to such a view. First, it would be, even from the strictly objective point of view, the point of view of physical science, to construe the thing mind entirely in terms of the behaviour of its stages antecedent to the present; that is, entirely in terms of descriptive content, by use of the categories of retrospective interpretation. And, second, it does not follow that because a mental way of regarding the world is itself a genetic growth, therefore its meaning is exhausted in the conditions of its genesis. Let us look at these two points a little more in detail.

1. A chemist seems justified in looking upon atmospheric air as explained by the formula for a mixture of nitrogen and hydrogen, for the reason, and this is his practical test, that the behaviour of air confirms that view. His confidence in his statements of history can only be justified on the ground that present history never contradicts it. But as

soon as a new experiment showed that new behaviour may be different, and may contradict the reports of history, he looks for a new thing, argon — new in the sense, of course, that the historical manifestations of the kind of reality in what is called air had never before brought it to recognition. In other words, the nature of air had been stated in terms of oxygen and nitrogen ; but he now sees that the statement, founded on what was known of origin — and that is what origin means in all these discussions — was inadequate. This would seem to admit, however, that if the problem of origin could be really exhausted, that of nature would be exhausted too; and no doubt it would. But it is a corollary from the second point of objection, soon to be made, that the problem of origin can never be exhausted, even by philosophy, without an appeal to other than the historical or retrospective categories.

But before we pass on to the second objection to the position that a thing which is admitted to have had a natural history must have its interpretation adequately given in that history, and that this applies also to the very categories by the use of which its denial is effected — before going farther we may note an extreme case of the main position as sometimes argued by evolutionists. If, it may be said, the mind has developed under constant stimulations from the external world, and if its progress consists essentially in the more and more adequate representation in consciousness of the relations already existing in the external world, then it follows that these internal representations can never do more than reflect the historical events of experience. Consciousness simply testifies again to the real as it has been testified to her before. How, then, can there be any such thing as a phase of reality — called the prospective — which is not subject to plain statement under natural law ? [1]

This is a very common criticism of all thoroughgoing statements of mental evolution. It rests on the mistaken view, just pointed out, that a statement of the historical career of a thing can ever be an adequate statement of its nature; in other words, that the origin of the categories of thought can tell what these categories will do — what their function and meaning is in the general movement of reality. Consciousness is entitled to a hearing in terms of its behaviour solely. The behaviour, attitudes, etc., represented by 'prospective' thought are there just as its behaviour represented by its history is there. Who would venture to say that consciousness of a relation in nature is in no sense a different mode of behaviour from the relation itself in nature? The real point is in what I have already tried to put in evidence: that such a construction involves the assumption that reality in its movement defines all her own changes in advance of their actual happening. The very series of changes which constitute the basis in experience for the growth in consciousness of the category of change are the basis also for the new aspects of reality (say consciousness)

---

[1] It is this supposed necessity that leads Mr. Huxley to hold that evolution cannot explain ethics, *i.e.*, the supposed necessity that the validity of ethical values must be adequately found in the terms of their origin ; for, says he, the pursuit of evil would have as much sanction as that of good, for both are in us, and they would have the same origin (*Evolution and Ethics*, esp. p. 31). But to say, as we do, that the appeal made by the word ' ought ' is a ' prospective ' appeal, as opposed to the description of the ' is,' which is ' retrospective,' does not require us to say that the impulse to recognize either is not a product of evolution. My discussion of Professor Royce's attempt (*International Journal of Ethics*, July, 1895) to show the psychological origin of the antithesis between ' ought ' and ' is,' may be referred to (*Ibid.*, October, 1895, now reprinted in the volume *Fragments in Philosophy and Science*, Scribners, 1902, pp. 70 ff.).

which are held to be only a putting in evidence of the relations already existing in nature. If consciousness is no new thing—on our behaviour definition of thing—then knowledge of the historical movement of reality must be not at all different from the movement which has led up to the knowledge. The discovery of the principle of evolution, for example, is not a new event added to the fact that the series evolving was there to be discovered !

But we may be even more concrete. The writer has developed a view of mental development which not only makes each stage of it a matter of legitimate natural history, but goes on to say that the one process of motor adjustment is imitative in type. What could be a more inviting field for the criticism : imitation is mere repetition. How can anything new come out of imitation ? Not only is consciousness merely repeating the relationships already present in nature, but the development of consciousness itself is merely a series of repetitions of its own acts. This criticism has already been made, especially with reference to volition. How, it is asked, can anything new be willed if volition is in its origin only imitation become complex ?

The reply serves to make concrete what has been said immediately above. The counter question may be put: why cannot anything new come out of imitations ? Why may not the very repetition be the new thing, or the condition of it ? To deny it is to say that by looking at the former instance, the historical, after its occurrence, you can say that that occurrence fully expressed mental behaviour. On the contrary, the prospective reference gained by the imitation may bring out something new ; the repetition may be just what is needed to develop an important stage in the career of mental reality. In itself, indeed, an imi-

tation is no more open to the objection we are considering than any other kind of mental behaviour, and it is not
allowed that imitation is no more than repetition, — though,
of course, in certain cases it may be no more, — but it
seems to be open especially to this criticism because it
emphasizes the very point that the current objection to
natural history hits upon, *i.e.*, that it makes the mind only
a means of reinstatement of relations already existing in
nature, and then makes imitative repetition the explicit
method of mental history.

§ 8. *The 'Intuition' View*

2. The second answer to the view now being criticised
may be put in some such way as this. It does not follow
that because a product — one of the categories of organization, such as design, the ethical, etc. — is itself a matter of
gradual growth, its application to reality is in any way
invalidated. A category must be complete, ready-made,
universal, without exceptions, we are told, in order that its
application to particular instances be justified. But I fail
to see the peculiar and mysterious validity supposed to
attach to an intuition because whenever we think by it
we allow no exceptions. Modern critiques of belief and
modern theories of nervous habit have given us reasons
enough for discarding such touchstones as 'universality'
and 'necessity.' And modern investigations into the
race development of beliefs have told us how much better
an aspect of reality really is because at one time people insisted in thinking in a certain 'intuitive' way about it. The
whole trouble, as I think, with the intuitional way of thinking is curiously enough that fallacy which I have pointed
out as being a favourite one of the evolutionists. The evolu-

tionists say that an intuition is of no value when construed prospectively, *i.e.*, as applying to what 'must be' beyond 'what is'; it gets all its content, and all its force, from experience. Therefore, all reality is to be construed retrospectively, and no 'thing' is possible except as accounted for as an evolution from historical elements. True, after things have happened — it nevertheless fails by thinking career all finished. Why may not experience produce in us a category whose meaning is prophetic?

On the other hand, the intuitionists oppose the evolutionists in this way, saying: no thing is possible except as in some way evidenced for. The intuitions are universal and necessary. As such their evidence cannot be found in experience. To admit that they had developed would be to admit that their evidence could be found in experience. Consequently they carry their own evidence, and their own witness is all the evidence they have. The fallacy again is just the assumption that reality is finished; that categories of retrospective reference exhaust the case; that the series of events which are sufficient ground for the origin of the category might also be sufficient evidence of its validity; that there is a sharp contradiction, therefore, between a doctrine of derivation from experience (which is inadequate as evidence) and application beyond experience. But when we come to see that the categories of prospective thought are equally entitled to application with those of retrospect, we destroy the weapon of evolution to hurt the validity of mental utterances, but at the same time we knock out the props upon which the intuitionist has rested his case.

The case stands with mental facts, to sum up, just about as it does with all other facts. An event in nature stays what it is until it changes. So with an event or a belief or any other thing in the mind of the race. It stays what it is until it has to change. Its change, however, is just as much an element in reality as lack of change is; and the weakening of a belief like any other change is the introduction of new phases of reality. A doctrine which holds to intuitions which admit of no prospective exceptions, no novelties, seems to me to commit suicide by handing the whole case over to a mechanical philosophy; for it admits that all validity whatever must be cut from cloth woven out of the historical and descriptive sequences of the mind's origin.

Our conclusions so far may be summed up tentatively in certain propositions as follows:—

1. All statements of the nature of a 'thing' get their matter mainly from the processes which they have been known to pass through; that is, statements of nature are largely statements of origin.

2. The statements of origin, however, never exhaust the reality of a thing; since no statements can be the entire truth of the experiences which they state unless they construe the reality not only as a thing which has had a career, but also as one which is about to have a further career; for the expectation of the future career rests upon the same historical series as the belief in the past career.

3. All attempts to rule out prospective organization or teleology from the world would be fatal to natural science, which has arisen by provisional interpretations of just this kind of organization: and also to the historical interpretation of the world found in the evolution hypothesis; for the category of teleology is but the prospective reading of

the same series which, when read retrospectively, we call evolution.

4. The fact that anything — and more especially mental products, ideas, etc. — has had a natural history, is no argument against its validity or worth as having application beyond the details of its own history; since, if so, then a natural history series could produce nothing new. But that is to deny the existence of the fact or idea itself, for it is a new thing in the series in which it arises.

All these points may be held together in a view which gives each mental content a twofold value in the active life. Each such content, by its function as a genetic factor in the progressive development of the individual, begets two attitudes. As far as it fulfils earlier habits it begets and confirms the historical or retrospective attitude; as far as it is not entirely exhausted in the channels of habit, so far it begets the expectant or prospective attitude.

§ 9. *The Meaning of the Category of Causation*

There are one or two points among many suggested by the foregoing which it may be well to refer to — selected because uppermost in the writer's mind. It will be remembered that in speaking of the categories of organization as having prospective reference, I adduced instances largely drawn from the phenomena of life and mind, contrasting them somewhat strongly with those of chemistry, physics, etc. The use afterward made of these categories now warrants us in turning upon that distinction, in order to see whether our main results hold for the aspects of reality with which these other sciences deal as well. It was intimated above in passing that other categories of reality, such as causation and mechanism, are really capable of a similar evaluation as that given to teleology. This possibility may now be put in a little stronger light.

It is evident, when we come to think of it, that all organization in the world must rest ultimately on the same basis; and the recognition of this is the strength of thoroughgoing naturalism and of absolute idealism alike. The justification of the view is to be made out, it seems to me, by detailed investigation of the genetic development of the categories. The way the child reaches his notion of causation, for example, or that of personality, is evidence of the way we are to consider the great corresponding race categories of thought to have been reached; and the category of causation is, equally with that of personality or that of design, a category of organization. The reason that causation is considered a cast-iron thing, implicit in nature in the form of 'conservation of energy,' is that in the growth of the rubrics of thought certain great differentiations have been made in experience according to observed aspects of behaviour; and those events which exhibit the more definite, invariable aspects of behaviour have been put aside by themselves; not of course by a conscious convention of man's, but by the conventions of the organism working under the very method which we come — when we make it consciously conventional — to call this very category of organization. What is conservation but a kind of organization looked at retrospectively and conventionally? Does it not hold simply because my organism has made the convention that only that class of experiences which are 'objective' and regular and habitual to me shall be treated together, and so shall give rise to such a regular mental construction on my part?

u

But the tendency to make all experience liable to this kind of causation is an attempt to undo nature's convention — to accept one of her results, which exists only in view of a certain differentiation of the aspects of reality, and apply this universally, to the subversion of the very differentiation on the basis of which it has arisen. The fact that there is a class of experiences whose behaviour issues in such a purely historical statement and arouses in me such a purely habitual attitude, is itself witness to a larger organization — that of the richer consciousness of expectation, volition, prophecy. Otherwise conservation could never have been given abstract statement in thought.

The reason that the category of causation has assumed its show of importance, is just that which intuitionist thinkers urge; and another historical example of confusion due to their use of it may be used for illustration. Causation is about as universal a thing — in its application to certain aspects of reality — as could be desired. And we find thinkers of this school using this fact to reach a certain statement of theism. But they then find a category of 'freedom' claiming the dignity of an intuition also; and although this comes directly in conflict with the universality ascribed to the other, nevertheless it also is used to support the same theistic conclusion. The two arguments read: (1) an intelligent God exists because the intelligence in the world must have an adequate cause, and (2) an intelligent God exists because the consciousness of freedom is sufficient evidence of a self-active principle in the world, which is not caused. All we have to say, in order to avoid the difficulty, is that any mental fact is an 'intuition' in reference only to its own content of experience. Intelligence viewed as a natural fact, *i.e.*, retrospectively, has a cause; but freedom in its meaning in reality, *i.e.*, with its prospective outlook, is prophetic of novelties — is not adequately construed in terms of history. So both can be held to be valid, but only by denying universality to both 'intuitions,' and confining each to its sphere and peculiar reference in the make-up of reality.

§ 10. *Definition of 'Origin'*

Another thing to be referred to in this rough discussion concerns the more precise definition of 'origin.' How much of a thing's career belongs to its origin? How far back must we go to come to origin?

Up to this point I have used the word with a meaning which is very wide. Without trying to find a division of a thing's behaviour into the present of it as distinguished from its history, I have rather distinguished the two attitudes of mind engendered by the contemplation of a thing, *i.e.*, the 'retrospective' attitude and the 'prospective' attitude. When we come to ask for any real division between origin and present existence we have to ask what a thing's present value is. In answer to that we must say that its present value resides very largely in what we expect it to do; and then it occurs to us that what we expect it to do is no more or less than what it has before done. So our idea of what is, as was said above, gets its content from what has been — which is to inquire into its history, or to ask for a fuller or less full statement of its origin or career. So the question before us seems to resolve itself into the task of finding somewhere in a thing's history a line which divides its career up to the present into two parts: one properly described as origin, and the other not. Now, on the view of the naturalist pure and simple, there can be no such line. For the attempt to construe a thing entirely in terms of history, entirely in the retrospective categories, would make it impossible for him to stop at any point and say 'this far back is nature and further back is origin'; for at that point the question might be asked of him: 'what is the content of the career which describes the thing's origin?' —and he would have to reply in exactly the same way that he did if we asked him the same question regarding the thing's nature at that point. He would have to say that the origin of the thing observed later was described by career up to that point; and is not that exactly the reply he would give if we asked him what the thing was which then was? So to get any reply to the question of the origin of one thing different from that to the question of the nature of an earlier thing, he would have to go still farther back. But this would only repeat his difficulty. So he would never be able to distinguish between origin and nature except as different terms for describing different sections of one continuous series of aspects of behaviour.

This dilemma holds also, I think, in the case of the intuitionist. For as far as he denies the natural history view of origins and so escapes the development above, he holds to special creation by an intelligent Deity; but to get content to his thought of Deity he resorts to what he knows of mental behaviour. The nature of mind then supplies the thought of the origin of mind.

To those who do not shut themselves up, however, to the construction of things in the categories of realized fact, of history, of 'retrospect,' the question of origin is a fruitful one apart from the statement of nature. For at any stage in the career of a thing the two methods of thought are equally applicable. When we ask how a thing originated, we transport ourselves back to a point in its career at which the 'prospective' categories got a filling not *at that stage* already expressed in the content of history. The overplus of behaviour is said to have its origin then, even though afterward the outcome be statable in the categories of retrospect which have *then been widened by this event.* For example, volition originates in the child at the point of its life at which certain conscious experiences issue out of old content — experiences which were not previously present, to the child, whatever other complications of content were. But once arisen, the experience can be construed as a continuation of the series of events which make up mental history. To the positivist and to the intuitionist a sensational account of the genesis of volition, and to the intellectual idealist an ideological account of it, rule volition out of reality just by the fallacy of thinking exclusively in retrospect. But in truth we should say : granted either account of its origin, it leaves philosophy still to construe it ; for if we estimate volition from facts true before volition arose, the sources do not fully describe it ; and if we wait to view it after it arises, then the full statement of career must include the widened aspects of behaviour which the facts of volition afford.[1]

§ 11. *What is Potentiality?*

It is interesting also to note, as another case of application of this general distinction between the mental habits represented respectively by the terms 'prospective' and

---

[1] In the *Psychological Review* for September, 1895 (reprinted in *Fragments in Philosophy and Science*, IV.), I have criticised the idealists' view that the Absolute can be exhausted by our thought, *i.e.*, can be adequately expressed in terms of the organizations of content already effected.

'retrospective,' that it gives us some suggestions concerning the very obscure concept called potency or 'potentiality.' This soi-disant concept or notion has been used by almost every conceivable shade of thought as the repository of that which is unexplained.   Aristotle started the pursuit of this notion and used it in a way which shed much light, it is true, upon the questions of philosophy concerned with change and organization; but his failure to give any analysis of the concept itself has been an example ever since to lesser men.   It is astonishing that, with all the metaphysics of causation which the history of philosophy shows, there has been — that is, to my knowledge — no thoroughgoing attempt to trace the psychological meaning of potentiality. How common it is to hear the expression, 'this thing exists, not actually, but potentially,' given as the end of debate — and accepted, too, as the end.   I do not care to go now into a historical note on the doctrine of potentiality; it would be indeed mainly an exposition of a chapter of Aristotle's metaphysics with the refinements on Aristotle due to the logic of the schoolmen and the dogmatics of modern theology.   It may suffice to say something of the natural history of the distinction between potential and real existence in the light of the positions taken above.

In brief, then, as we have seen, there are two aspects under which reality must in all cases be viewed, — the prospective and the retrospective.   The retrospective, as has been said, is the summing up of the history which gives positive content to the notion of a thing considered as accomplished career.   This aspect, it seems clear, is what we have in view when we speak of 'real' in contrast with 'potential' existence.   It is not, indeed, adequately rendered by the content supplied by retrospect, since the fact that

the two predicates are held in mind together as both together applicable to any concrete developing thing, forbids
us to construe real existence altogether apart from the fact
that it has a further issue in later career. It is a great
merit of Aristotle that he forbade just this attempt to consider the *energeia* apart from the *dunamis*. But, nevertheless, it is true psychologically that real existence as a con-
tent-predicate is exhausted by the survey of the backward
aspect of the series of changes which give body to reality.

And it seems also evident at first blush that potential
existence is equally concerned with the prospective reference of the thought of things. That this is so is perhaps
the one element in the notion of potency that all who use
the word would agree upon. But this is inadequate as a
description of the category of potentiality. For if that
were all, how would it differ from any other thought of the
prospective? We may think of the future career of a thing
simply in terms of time; that, we would probably agree,
does not involve potentiality. A particular potency is confined to a particular thing, *i.e.*, to a particular series of
events making up a more or less isolated career. If only
the bare fact of futurity were involved, why should not any
new unrolling of career be the potency of anything indis-
criminately?

This leads us to see that potency or potentiality, even
when used in the abstract, is never free from its concrete
reference. And this concrete reference is not that of conception in general, only or mainly; the concrete reference
of conception generally is a matter of retrospect, *i.e.*, of the
application of the concept to individual things, as far as
such application has been justified by historical instances.
Indeed, it is the very occurrence of the historical instances

which has given rise to the concept, and it generalizes them.

So when we put ourselves at the point of view of the concrete, we have to ask what is actually meant by us when we say a thing exists potentially, over and above the mere meaning that the thing is to exist in the future. We have seen that one added element of meaning is that the thing which is to exist in the future is in some way tied down in its manifestations to something that already exists actually ; it must be the potentiality of some one thing in order to be a potentiality at all. Now, what more can it be ?

Of course the ordinary answer is at once on our lips : the answer that the bond between the thing that is and the thing that is to be is the bond of causation. The potentiality is the unexpressed causal 'efficacy' of the thing that is. But when we come to ask what this means, we find that we are hiding behind one of the screens of common sense. The very fact of cause, whatever bond it may represent from an ontological point of view, is at least a fact of career. The effect is a further statement of the career of the thing called the cause. Now, to say that the potency of a thing is its unexpressed causal power, is only to say that the thing has not finished its career, and that is a part of the notion of a thing in general. That fact alone does not in any way define the future career for us, except in the way of repetition of past career. We merely expect the thing to do what it has done before, not to become some new thing out of the old. In short, the category of causation is not adequate, since it construes all career retrospectively.

We have, therefore, two positions so far, finding (1) that every potency is the potency of a thing, and this means that it gets its content in some way from the historical

series which that thing embodies ; but (2) that it is something more than a restatement of any or all of the elements
of the series thus embodied.   Now, what else is there ?

The remaining element in the category of potentiality
involves, it seems, a very subtle movement of the mind along
the same distinction of the prospective from the retrospective.   Briefly, the potentiality which I ascribe to a thing is
my general expectation of more career in connection with it,
with the added sense, based on the combined experiences
of mine that the prospective does get a retrospective filling
after it has happened, that the new career of the thing to
which I ascribe the potency, although not yet unfolded,
will likewise be capable of retrospective interpretation as
further statement of the one series which now defines the
thing.

In short, there are three elements or phases of consciousness involved : first, let us say, the general prospective
element, the expectation that something will happen ; second, the causation or retrospective element, the expectation
that when it has happened it will be a consistent part of
the history of the thing; and, third, the conscious setting
back of my observation to the dividing line between these
two points of view, and the contemplation of the thing
under both of them — both as a present thing, and as a
thing for what it will be when the future becomes present.

For example : I say that a tree expresses the potency or
potentiality of the seed.   This means three very concrete
things.   I expect the seed to have a future ; I expect the
future to be a tree — that is, a thing whose descriptive
series is continuous with that already descriptive of the
seed, — and, finally, I look upon the seed as now embodying
the whole tree series thus artificially present in my thought.

§ 12. *The Origin of the Universe; Further Problems*

On the view developed in this paper, the question of the ultimate origin of the universe may still come up for answer. Can there be an ultimate stopping-place anywhere in the career of the thing-world as a whole? Does not our position make it necessary that at any such stopping-place there should be some kind of filling drawn from yet antecedent history to give our statement of the conditions of origin any distinguishing character? It seems to me so. To say the contrary would be to do in favour of the prospective categories what we have been denying the right of the naturalist to do in favour of those of retrospect. Neither can proceed without the other. The only way to treat the problem of ultimate origin is not to ask it as an isolated problem. Lotze says that the problem of philosophy is to require what reality is, not how it is made; and this will do if we remember that we must exhaust the empirical 'how' to get a notion of the empirical 'what,' and that there still remains over the 'prospect' which the same author has hit off in his famous saying, 'Reality is richer than thought.' To desiderate a what which has no how — this seems as contradictory as to ask for a how in terms of what is not. It is really this last chase of the 'how' that Lotze deprecates — and rightly.

Certain further applications:[1] to the discussion of *freedom;* to the discussion of *ideals;* criticism of the general concept of *law* from this point of view; applications in

---

[1] Questions suggested to the members of the Psychological Seminary for discussion. A further development of the point of view of this paper by one of the members of the Seminary, Professor W. M. Urban, is to be found in the *Psychological Review*, January, 1896, pp. 73 ff.

ethics (cf. with Royce's distinction of 'world of descrip-tion' from 'world of appreciation'); the question of the notion of time (*i.e.*, is the distinction between the pro-spective' and 'retrospective' merely one of time, or does the notion of time find its genesis in this difference of mental attitude?); the problem of value (are all values prospective?).

CHAPTER XIX

The Theory of Genetic Modes

On the basis of the conclusions of the preceding chapter we may take up a question which concerns the method of positive science and the nature of the formulations which science is able to make. If it be of the nature of all 'things' that they are in process of change, and if the growth of experience be such that two aspects of reality alike engender mental attitudes, called respectively the 'prospective' and the 'retrospective,' then it becomes of great importance to determine, so far as may be, *the relation of the mind to its objects, in the body of knowledge called science.* There are two general positions, held more or less explicitly by different writers, with reference to which the following discussion may be conducted.

§ 1. *Agenetic Science*

In the first place, the processes or events with which science deals may be considered under certain mental rules or conditions, which represent an ideal of regularity in a series of transformations which run their course in a finished and traceable form. The 'shorthand' descriptions of such processes state the 'laws' which, if these ideals or rules be conformed to, phenomena, broken in upon — cut in cross-section, as it were — at any point of their development for purposes of observation, will be found to illustrate. An adjunct to this method is the further procedure of so arranging the conditions that the phenomena are caught going through certain of the more recondite phases of their behaviour; this last is called experimentation.

Such is the method of the 'physical' sciences — physics and chemistry — as distinguished from the 'natural' or biological sciences. The postulates of this procedure are (1) *uniformity* — which means no more nor less than 'agenetic'[1] regularity, or the absence of any sort of change which is not exhaustively interpreted in terms of preceding change of the same order. With this there is (2) the postulate of some sort of *lawfulness* — the requirement that natural phenomena be not capricious in their behaviour, but that experience so order itself by law that illustrations of what the law means, or what it has come to mean on the basis of just these experiences, may actually and at any time be found. As representing one way of looking at science this 'agenetic' point of view is made extreme in the claim that this procedure, which tacitly fails to recognize the genetic, or which explicitly confines itself to the 'agenetic,' is the exclusive procedure and exhausts the resources of science.

Such a view, which I shall henceforth call the 'agenetic theory'[2] of science, rests upon certain interesting and important mental movements. If we hold that the growth of experience, whereby it reaches maturity in what we call 'thought,' is by the formation of certain categories or habits, then it seems necessary to say that, so far as experience is organized at all, it must be in these categories; and further, that a category itself reflects something

[1] A term meaning, of course, not genetic, as genetic is explained below.
[2] Positively it is the point of view of ' quantitative ' or exact science.

of the uniformity and lawfulness of experience. But we saw on an earlier page that it is not necessary that the categories, which are themselves the outcome of regular experience, should apply only to phenomena which themselves illustrate that regularity. There are certain categories of thinking and of objective interpretation whose content is the changing, the genetic, the in-a-sense-capricious, yet which themselves stand for and represent *in mental growth* the uniformity and lawfulness of experience. So it becomes necessary to distinguish between those types of experience which illustrate a mental rule or category on the one hand, and those which produce it on the other hand. It may be quite true that one cannot think of a change as taking place in nature without asking for the changes which preceded it; this is the requirement that the category of change finds in phenomena its justification; but it is quite a different thing to say that the antecedent change which this category of thought postulates is a sufficient statement of that which follows, and that for which a scientific account is sought. There are categories, therefore, whose application requires change or variation even in the midst of the regularities by which they themselves are produced.[1] This it is which characterizes the 'genetic' categories.[2]

§ 2. *The First Postulate of the Theory of Genetic Modes*

So important is this consideration for a criticism of science, that the failure to recognize it constitutes a

---

[1] The category of change, indeed, is constituted by the *regularity of change*.

[2] The phrase 'dynamic categories' is sometimes used (see Ormond, *The Foundations of Knowledge*, Part II., Chap. VIII.), but with a meaning not in all respects coincident with that here given to the term 'genetic.'

vitiating element in most attempts to construct a scientific view of the world. In the language of our earlier distinction, they make the retrospective exhaustive, and use only static formulas for the phenomena which are essentially genetic, prospective, and dynamic. This procedure employs a mental shorthand which is correct so far as it goes, and which is quite right in its demand that phenomena, to be natural at all, shall fulfil its statements; but it fails to recognize the possibility that *these same phenomena may be yet more* — may fulfil requirements of a genetic sort which such formulas do not construe nor recognize.

This outcome it is which I wish to set down as *the first or negative postulate of what is here called the 'theory of genetic modes.'* This postulate may be stated as follows: *the logic of genesis is not expressed in convertible propositions.* Genetically $A = B$; but it does not follow that $B = A$. In its material application this takes on two forms: first, if xy is invariably followed by z, it does not follow (1) that z is invariably preceded by xy, nor (2) that nothing more than z invariably arises subsequently to xy. In the language of chemistry these two points read: granted that oxygen and hydrogen produce water, it does not follow (1) that water may not be produced by something else than oxygen and hydrogen, nor (2) that if the water be reduced to oxygen and hydrogen again, something else than water may not have been produced and again destroyed along with the water.

This, some one may object, traduces the law of cause and effect as generalized in the formula for the conservation of energy. Not so; but it does traduce certain illegitimate extensions of that law. It says explicitly that there are certain aspects of phenomena which that law — admitting the postulates of uniformity and lawfulness mentioned above — has the right to construe, and which we are bound to recognize when we use the categories which experience of these aspects has engendered. But it does not work negatively or conversely; it cannot dictate to reality its future working, nor say that in the very experiences so formulated there may not be more than these formulations get out of them.

For an illustration of this point, let us go direct to a critical case. Brain changes are accompanied, say, by acts of conscious volition. If we saw only the outside of a man's brain, our science of brain changes — the shorthand description of what we see — would seem to exhaust the phenomena; but all the while there would be present, inside the man's head in some sense, and escaping our description altogether, the phenomena of volition. Now suppose that these inner phenomena of volition are present only at a certain stage in the development of a series of brain changes, appearing when the individual is from six to nine months old. Admitting for the moment that the description made from the outside is exhaustive, both earlier and also later on in the series, the later terms simply being further along and perhaps more involved; yet this gives no inkling of the change from one form or mode of consciousness to the other — that is, of the rise of volition. All that another science, psychology, takes cognizance of — the mental transformations — are additional things in the world, aspects of reality not in so far touched by the formulations of quantitative science. Who can tell, indeed, what modes of existence may come and go with the development of changes in the brain?

§ 3. *Genetic Modes*

But, in fact, we cannot admit the assumption just made, that quantitative science is exhaustive even for the brain changes taken alone — to bring out a point which takes us further, and which may seem still more out of touch with the claims of physical science. I contend that absolutely new and unheard-of phases of reality may 'arise and shine' at any moment *in any natural series of events — constituting new 'genetic modes.'* Considering the origin and nature of the categories of thought, whatever our theory of the method of their genesis may be, we find that they are modes of function *selected for their utility* as furnishing interpretations of experience.[1] It is evident, then, that it is impossible to discount or deny, by their use, any modes of existence or reality *which they do not interpret.* As is intimated on an earlier page, animals of different grades may have such varying sense-organs and such varying qualities of sensation, feeling, or other mode of consciousness, as to make their apprehension of the world of bionomic changes very different one from another. To a creature in which the olfactory lobe is developed in a preponderating way, smell may be the control sense, and interpretations by smell may be the final tests of what to this creature are the realities of his life; to another, touch, to another, vision, may be the leading sense. Now each, in his several sphere, must think, must, in general, psychologize, under his own rubric; each has his test of truth. And he must also in so far legislate it as final upon experience. But yet, other animals may have other measures, tests, interpretations, — other realities of which

---

[1] This is the general outcome of Chap. XVII., on 'Selective Thinking.'

x

he knows nothing. The very origin of the categories which we use in science restricts their application, since there may be other types of experience which are so far untouched and which might be construed only under other categories.

This becomes more evident as we rise in the scale of the sciences, because the relativities of apprehension are ever increasing with mental advance. As I have endeavoured to show in another place, the evolution of the higher faculties is by adaptation to a system of environmental relationships. The relation of the individual to this system is, as evolution proceeds, increasingly remote and indirect.

Memory arises as an adaptation to the distant in time, and as a weapon of prophecy, to the distant in space.[1] Imagination and thinking[2] are modes of psychic process which deal with generalized, abstract, not-fully-present data; and in so far as the data are not fully present in so far the relativeness of the result is increased. The child acts upon his sense of the general, and constantly finds that it fails in reference to the particular. He is ever readjusting himself with reference to conditions with which he has already coped with more or less success, but without finality. It would seem to be only the fixed, the strictly organic functions, which minister to his progress by immediate contacts with the brutely concrete and bruising things of time and space, which really 'hold' fast and inflexible for us. Other accommodations are, by their nature as accommodations, parts always of a growing system, elements of a genetic process, factors of a larger accommodation

---

[1] Cf. the volume *Mental Development*, Chap. X.
[2] *Ibid.*, Chaps. X., XI.

yet to be achieved. And it is plain that this must be so. The congenital, whether organic or mental, is a variation, selected just by reason of its close-fitting character upon this fact, relation, or need in life ; while the other characters — the plastic, mobile, intelligent — have their chance and their utility only in the shifting, change-exhibiting sorts of experience to which the genetic growth process must conform, but which it can never really exhaust. This distinction reflects itself in the entire system of mental accommodations — what is called above the ' determination of thought,' — in an aspect of mental growth, a general attitude which in so far directly antagonizes the fixities of the congenital and immediate, and holds a brief for relative truth, relative life, relative right — since the world itself as a whole, by being itself a world of change and growth, is a system of relative parts.

Consciousness, therefore, not only accepts the old, by those adaptations by which it categorizes the familiar ; it also finds the new and welcomes it. In the accommodations to the social environment and to tradition through which the consciousness of self makes what progress it does into this stage or that, an ideal arises to embody just this consciousness of the relativity of all possible concrete determinations of mental content or conduct. Were reality fixed and were adaptation ended, ideals would be impossible. Whence the thought of progress toward the better, the more fit — in whatever sphere, — if all were now attained, and the future had no largess, no rewards, no unexplored tracts, no new realities to confront and possibly to subdue us ? We cope with the new, indeed, by this tentative outreach toward it, armed with our categories of description and interpretation. In so far as these are adequate, they reflect earlier stages in the unfolding of the same system.　But the 'arming' is inadequate for full interpretation, since it is forged in the fires of the past. The ideals, the values yet in process, and always to be in process, of achievement, get their impelling power from the very experience that knowledge and life are functions of a genetic process of which our formulated realities are passing phases.

This might be carried out in a philosophical view of reality, — a theoretical doctrine of metaphysics, — but that is not my intention here.　The only safe course for science, however, is to recognize these things.　Genetic science is competent to make the reservation always, in the presence of each of the applications and explanations of exact and numerical science, that *it is a cross-section, not a longitudinal section, to which the quantitative and analytical formulas apply;* or that, if they apply throughout a serial process, — as in a series of successive transformations of energy, — *that* is proof that the process in that case *is not a genetic one.*　It is the genetic aspect, we must hold in such cases, which has escaped the formula; the success of the quantitative and analytic methods is itself the evidence that no really genetic movement has occurred out of the natural aspects of things; in other words, only those have been taken which illustrate the repetitions, not the adaptations, of nature.

§ 4.　*Genetic Science*

We may undertake, in view of these considerations, to state the actual relation which we are justified in holding to subsist between exact or agenetic science, so-called, on the one hand, and genetic science, on the other hand.

This leads us to the second great class of views which are possible regarding the province of knowledge and the relation of mind to nature. I say class of views, since it is a class, in which many varied constructions in detail might be worked out. So far as the view which follows has details, that is, attempts to apply the line of distinctions now made to the actual relations of the sciences, they may be taken as my personal views, and they should not be allowed to prejudice the truth of the general distinction itself.

Starting out with the development of the preceding chapter and adding the further thoughts stated on the pages immediately above, we have a certain way of construing science, which allows full sweep to the genetic point of view. All knowledge is in its essence, as cognition, retrospective. As Kant claimed, knowledge is a process of categorizing, and to know a thing is to say that it illustrates or stimulates, or functions as, a category. But a category is a mental habit; that is all a category can be allowed to be — a habit broadly defined as a disposition, whether congenital or acquired, to act upon, or to treat, items of any sort in certain general ways. These habits or categories arise either from actual accommodations with 'functional' or some other form of utility selection, or by natural endowment secured by selection from variations. Organic selection affects the parallelism between these two lines of origin, in the way pointed out in the earlier pages of this work.

In dealing with any set of data or phenomena the question comes up as to what categories apply — what habits of treatment are brought out and illustrated when we get all we can out of these facts.

Invoking the shades of the Old Masters of Greece, we think with them of the antithesis between being and becoming. We ask of this and of that — of everything, indeed — not only what its value in fact, but what its worth in prospect; not only for its place in the has-been, but for its claim on the yet-to-be. We cannot explain it, even in its network of shifting observed relations, without projecting out before us and before it an expected career. This is the distinction made above between the retrospective and the prospective point of view. The application of it here is to the theory of objects, as such. We must treat the yet-to-be of the object as being as real as the yet-to-experience of the mind. The object is an object for cognition when it is a substantive, a term, in a network of relationships — as it were, a knot trailing its 'fringe' before and after.[1] The explanations of exact science, which analyze it into those elements only which went into its composition, tie up the fringes that trail behind, and so make a series of knots extending far back into the dim distance of time, of history, and of logic. But the fringes which *stretch out before* — these fly free in the wind; and while no continuation of the threads is to be seen, and no knots of further knowledge can yet be tied, still we have the assurance that *these do not break where they seem to end*, an assurance as indubitable and as well guaranteed in our mental constitution as our assurance of the continuity of the back-leading threads already tied up in knots by the formulas of exact science.

This we know because, by waiting, we find out always that this is the outcome. Never has this expectation failed. And it cannot fail; for with it would fail also our

---

[1] A figure made familiar in another context by William James.

trust in the retrospective formulas — the tying of past threads in knots. For the event now present rides by us and becomes past; and the very threads we assayed to trace with pains and failure, become those which form the backward fringe, and constitute history. The whole forms a chain. Experience is continuous. Our discoveries that events now gone, experiences now no more than memories, still fit into what we call the categories of knowledge — these discoveries are no more valid, from the point of view of genesis, than are the expectations and prophecies, which we perforce must also indulge, respecting the future which issues from the present.

So there is a genetic science, as there is a prospective attitude — a science of development and evolution. It is of the knowledge series which we are not able to read both ways, or which, if read both ways, has for each a different formula, that the term genetic is properly used.

§ 5. *The Second Postulate of the Theory of Genetic Modes*

We may write down, accordingly, as the second or positive postulate of the 'theory of genetic modes,' that *that series of events only is truly genetic which cannot be constructed before it has happened, and which cannot be exhausted by reading backwards after it has happened.*

To be sure we often apply the term genetic to all cases in which history is involved; cases in which there is a regular series of changes. But there are several cases of this.

If a series of events so exhausts itself that we may begin it over again and find the terms one by one again following their aforetime sequence, then this is not truly genesis; here instead is a static cycle, with a formula for recurrence or repetition, not for growth.

Again, we may find such a recurrent cycle of terms, but, besides, a something over which we clandestinely or overtly neglect. This neglecting is often explicitly done, notably in biology.

And yet again, we may come upon a condition of such complexity that the forces at play cannot be separated out one from another. I wish to make special mention of certain instances, especially of the sort mentioned second just above, which bring out the point of view of the theory of genetic modes.

The case of the recurring series is, in so far as it is a series, and not a mere term that recurs, a case in which the genetic may enter; for the question of growth may be asked of changes inside the series itself, and we may find that the terms as such are not recurrent, but represent an irreversible order; for example, certain series of changes of a chemical nature seem to be such. Of course it is the aim of exact science to reduce these series to those of the strictly repetitive type. A great instance of such reduction was the discovery of the law of gravitation, by which whole sets of unexplained serial phenomena were found to illustrate the repeated operation of attraction by the law of inverse squares. So, too, the reduction of the physical forces to terms of common work measured in energy, of which the quantity remains unimpaired. The reduction of all physical phenomena to such quantitative statements must remain the legitimate ideal of exact science. Yet while recognizing this, and recognizing the universal character of the category so exploited, we must at the same time make the reservation that even the thus-formulated facts may have, for all we know, other aspects also capable of formulation. Other shorthand expressions may be needed for their behaviour as parts of a larger whole which is constituted as the system which includes them is genetically unfolded.

§ 6. *History a Genetic Science*

History itself, considered as a science, illustrates the cases mentioned second and third above. There are various theories of history, yet all of them may be classed as in type falling under three headings. Those writers who reject the truly genetic from the sequences of history come first. In their theories they interpret history as a series of happenings under the law of cause and effect, showing from first to last a series of complications all of which may be considered as but different arrangements of given elements under the action of constant causes. This is strictly an attempt to make history a retrospective science, not only by the application of the categories of retrospect, but also by the claim that this application affords an exhaustive statement of possible knowledge of the series which comes to our apprehension in the events of days and years. There is nothing over — no meaning for higher interpretation than that formulated in the theory of the complication of elements under the law of causation.

A second view of history finds it practically lawless — a series of caprice-like discharges from the void. It is not an unfolding from anywhere to anything; but a series of terms whose sequence is absolutely unpredictable, because the terms are unrelated. This does violence to the categories of retrospect; and as I have said on an earlier page, any view that does that defeats itself, since it destroys the very lamp from which streams all our light, — not only the light of expectation, but also that of experience.

The third theory we may call, as the present writer has elsewhere called it,[1] the 'autonomic' theory. It holds — and it may therefore be used to illustrate the position developed here — that law must hold in history, since history is human experience; also that nowhere in its evolution does history, *after it has happened*, fail to fulfil the law of cause and effect, could we but unravel the intricacies of the phenomena; but that more than the categories of retrospect and of law are involved — *provided it be found out that they are*, that is, that more than conformity to this law *may be involved. History may be genetic;* to my view — though perhaps not necessarily to all forms of the autonomic view — it is. All history is sociology; it is also psychology; it is also ethics — it is all these, besides being, in a sense, biology and even physical geography. The autonomic view makes the claim simply that historical sequences shall afford their own interpretation. If there be a really genetic strain in the historical sequences, then it will appear. Each science that has the right, from the demarkation of its phenomena, to enter the field and to attempt to construct the historical material by its own shorthand formulas, shall have the fullest liberty to do so. *And each interpretation may be true.* Success is the only and the complete justification in each case.

Each of them may be true, because each of them

---

[1] *Dictionary of Philosophy and Psychology*, art. ' History,' *ad fin.*

may deal with an aspect which fulfils the demands of a certain sort of construction. To deny this in favour of an exclusive cause-and-effect theory is to violate our first postulate, as formulated above; it is to assert that retrospective formulations, *even when fully made out,* are by their own right exhaustive. In a discussion on another page, we may find an indication of how such double or multiple constructions of the same data may be possible—in the case of moral statistics. In individuals' actions, as seen, for example, in the statistics of suicide, the genetic character of the series is evident—a series of which each term is determined by an act of will, and illustrates a stage of mental progress, while yet statistics of the series, taken for a great many cases, are found to illustrate, in their distribution, the law of probabilities, as strictly as do the veriest mechanical events or the veriest 'chance' sequences.

Another case has also been discussed above, and is mentioned again below: that of biological evolution advancing under the law of natural selection, and at the same time possibly embodying purpose and teleology. Biological progress may be teleological, and really genetic—new stages of process, new genetic modes, appearing in the series—while, at the same time, the entire series, interpreted after it has happened, shows the character of regularity and uniformity which justify its construction in terms of natural selection from variations distributed in accordance with the probability curve.

§ 7. *The Biological Theory of History*

This general position may be given concreteness by a detailed case. It is evidently in antagonism to the view that human history can be exhaustively explained by the principles of organic evolution. This view has been recently stated with considerable force and dogmatism by Professor Karl Pearson in these words (*The Grammar of Science*, 2d ed.): "How far are the principles of natural selection to be applied to the historical evolution of man? History can never become science, can never be anything but a catalogue of facts rehearsed in more or less pleasing language, until these facts are seen to fall into sequences which can be briefly resumed in scientific formulæ. These formulæ can hardly be other than those which so effectually describe the relations of organic to organic phenomena in the earlier phases of their development. The growth of national and social life can give us the most wonderful insight into natural selection, and into the elimination of the unstable, on the widest and most impressive scale. Only when history is interpreted in the sense of *natural* history, does it pass from the sphere of narrative and become science. . . . In the early stages of civilization the physical environment and the more animal instincts of mankind are the dominating factors of evolution. Primitive history is not a history of individual men, nor of individual nations in the modern sense; it is the description of the growth of a typical social group of human beings under the influences of a definite physical environment, and of characteristic physiological instincts. Food, sex, geographical position, are the facts with which the scientific historian has to deal. These influences are just as strongly at work in more fully civilized societies, but their action is more difficult to trace, and is frequently obscured by the temporary action of individual men and individual groups. The obscurity only disappears when we deal with average results, long periods, and large areas. . . . Rivalry is at bottom the struggle for existence, which is still moulding the growth of nations ; but history, as it is now written, conceals, under the formal cloak of dynasties, wars, and foreign policies, those physical and physiological principles by which science will ultimately resume the development of man. Primitive history must be based upon a scientific investigation into the growth and relationship of the early forms of ownership and of marriage. It is only by such an investigation that we are able to show that the two great factors of evolution — the struggle for food and the instinct of sex — will suffice to resume the stages of social development. When we have learned to describe the sequences of primitive history in terms of physical and biological formulæ, then we shall hesitate less to dig deep down into our modern civilization and find its roots in the same appetites and instincts " (pp. 358–361 ; 362–363).

Such a view, if considered as an exhaustive account of history, rides rough-shod, I venture to think, over certain evident and vital distinctions. So much so that I place the objections in order, not, of course, taking space here for the repetition of the considerations on which they are based.

(1) Professor Pearson overlooks the distinction between what is intrinsic to a particular sort of organization, and that which merely conditions it, or is 'nomic' to it. In this case, it amounts to a failure to distinguish between the struggle for existence between groups and the inner organization of the group as a social whole. The former, ' group-selection,' is certainly a case of struggle for existence, but the main problem of the science of history and of sociology as such, is that of the forms and modes of

organization of the social relationships *within the group*.
Professor Pearson seems to see this later on where he
points out what he calls 'socialism,' which he makes the
'interest individuals have in organizing themselves owing
to the intense struggle which is ever waging between
society and society; this tendency to social organization,
always prominent in progressive communities, is a direct
outcome of the fundamental principle of evolution.'
Surely an easy way to solve the problem of social evolution! Is it because and in view of the 'intense struggle
between society and society' that social organization takes
place? This does not follow, even though we admit that
natural selection acts to preserve societies which are 'fit'
in this respect.

Would not a single social group on an island in the
Pacific sooner or later effect social organization and make
progress, provided they had the mental equipment?
Are there not certain characters intrinsically of a social
sort that make it possible — yes, necessary — for society
to exist? Can struggle and survival be a sufficient account of the actual evolution of English Economic History, let us say, of the rise and development of British
idealism, or of the evolution of republican principles
in France? History is a science principally of social
thoughts, ideals, psychological give-and-take, not mainly
of wars, considered as a form of struggle for existence,
which define and perpetuate the group-type in which
this or that social organization takes place, however
much importance we may give to the latter in its own
sphere.

(2) Professor Pearson fails to give any place to the
psychological factors, apart from such 'physiological

instincts' as desire for food and sex.[1] Truly a poverty-stricken list! Where is *thinking*, which even we selectionists must admit to be the prime utility of increasing nervous plasticity? Bagehot, writing long before Pearson, made much of 'group-selection,' but he saw its limitation, and signalized the 'age of discussion,' in which the controlling factor in a people's advance, the real key to their history, is their reasoning faculty. And Bagehot it was, as well, who pointed out the social process of *imitation* as one, at least, of the important agencies of socialization. This seems to illustrate what is said above, to the effect that the emphasis of natural selection as an all-sufficient principle has gone so far that it leads to the denial of the evident positive factors of endowment, variation, laws of change, etc., which are the essential motive principles of progress — in this case the psychological factors to which social progress is due — in favour of that merest shell of a truth, so far as social life is concerned, that like animals fight one another, and that the strongest lives to tell the story. Even as affecting the problem of group competition, what may we not say, for example, about the mental fact called *invention*?

Invention not only plays an extraordinary part in internal social organization and progress, — it escapes the barrier of heredity by a mighty bound, — but it serves to fit the competing group to survive. Suppose the knowledge of firearms and the use of smokeless gunpowder to be the possession of one only of two competing groups :

[1] One is reminded of Professor Pearson's own demonstration, in his statistical discussion (*Chances of Death*, Chap. I. p. 68) of the uneven distribution of families according to number of children, of the psychological interference with the normal birth-rate — an interference from a 'Malthusian restraint of population' exerted directly in opposition to the instinct of sex.

who can doubt the issue of their combat? Can we say then that the evolution which is determined by such a struggle is sufficiently explained by the statement of the strife between the two, with no allusion whatever to the firearms, or to the smokeless powder, or to the mental equipment that invented these? And shall we call this an explanation of history? It would seem, indeed, that we were bringing back 'home to roost' the charge which Professor Pearson makes against the historians, that they are merely cataloguing facts — and that his is, for all that, a very incomplete catalogue!

The case may indeed serve to give point to two of the main principles which it is the object of this work to set forth. First, if evolution is to take any account of facts, the psychological facts with the laws of their operation are not to be ignored. And if psychophysical evolution is to be the type which the true theory of evolution recognizes, then the correlations and dependencies of the two series of facts must be in all cases most carefully made out. Why, for example, select the craving for food, and not that for social companionship; why that of sex, and not that of religion? Professor Pearson speaks of biological principles as giving *natural* history, as though biology were in possession of a monopoly of nature. Surely the mind is a natural possession; and to say that imitation is a factor in social progress is as truly to recognize a natural history factor, as to say that struggle for existence is. The working of the mind in effecting an invention is every whit as natural a process as is the origin of variations by sexual reproduction. And second, it will not do to force the yoke of one science in this ruthless way upon the neck of another. Professor Pearson himself holds that science merely states shorthand formulas for the actual behaviour of phenomena; then let us look at actual phenomena, — social, historical, psychological, — and see how they work before we say that their formulas can 'hardly be other than' those of organic and inorganic phenomena. It is the attempt to reach positive rules for distinguishing one science from another, as we ascend in the hierarchy of knowledge, that is made in the theory of genetic modes.

I have put this criticism in a somewhat extreme form, no doubt, seeing that Professor Pearson does say that the socialistic instinct, as opposed to the individualistic, should have greater emphasis than is usually given it; but it is his principle that because the higher forms of endowment and organization have arisen under the operation of natural selection, that *therefore* the laws of their rise and progress in social and ethical life, history, etc., can be reduced to those of struggle for existence and natural selection; this, I contend, is mistaken. It is potent illustration of the denial of any possible genetic modes in the complex phenomena; it asserts that if we could master the conditions, we could not only predict future historical changes, but that the retrospective formulations of historical events stated in terms of biological law would be exhaustive of historical reality as such.[1]

---

[1] We cannot take up in this connection the more recent philosophical discussions of the science of history, although the 'theory of genetic modes' takes sides in the controversy. It says explicitly that history is capable of retrospective interpretation; but with equal explicitness, that such interpretation does not — or may not — exhaust the meaning of historical sequences. Each of these positions is denied by one party to the philosophical controversy, by the insistence either upon an exclusively 'scientific' or an exclusively 'humanistic' (for the most part *voluntaristic*) construction. An article summing up certain aspects of the controversy is that of Villa, 'Psychology and History,' *Monist*, XII., January, 1902, pp. 215 ff. (with literary citations).

Y

§ 8. *The Axioms of Genetic Science*

A survey of the sciences, according to the great divisions which are to-day current, serves to show that certain of the distinctions now suggested are fairly well recognized; but the most irreconcilable differences as to province, method, and preferential claim spring up about the lines of division, through the need of a principle which shall establish more exact boundaries. The general hierarchy of the sciences, starting with physics and chemistry, and passing up through the natural or biological, into the mental, and finally into the moral sciences, — this is well established. But we find the claim made, in conformity to the theory discussed on an earlier page, under the term 'agenetic science,' that the true method of science, and its one ideal, is the reduction of the complex phenomena of each of the higher, in turn, into statements of laws which hold for the lower, until we finally reach formulas which actually state all knowledge in the terms of the quantitative measurements of the physical and mathematical sciences.

Against such a demand and the scientific ideal which it erects, philosophical thinkers in certain branches of research have been in continual protest. And if what we have aimed to make out in our earlier pages be true, then this protest may be put in the form of a general distinction. The distinction holds as between each of the sciences and the one which lies below — the one upon which it depends in the way indicated by the term 'nomic.'[1]

We are able to say that what has been overlooked in each case, in the attempt to reduce a given sort of phe-

[1] Above, Chap. I. § 2.

nomena to lower terms, is the genetic aspect. It is the mistake of treating all phenomena by the method of 'cross-sections,' without supplementing such treatment with that involving the 'longitudinal section.' This is one self-repeating source of confusion. It fails to recognize the existence of genetic modes.

In a general and incomplete survey such as this, we may put in the following form the principles which we are justified in adopting, as axioms of the theory of genetic modes.

First, the phenomena of science at each higher level show a *form of synthesis* which is not accounted for by the formulations which are adequate for the phenomena of the next lower level. By 'lower' and 'higher' I mean *genetically before and after*, in the essential sense already explained.

Second, the formulations of any lower science are not invalidated in the next higher, even in cases in which new formulations are necessary for the formal synthesis which characterizes the genetic mode of the higher.

Third, the generalizations and classifications of each science, representing a particular genetic mode, are peculiar to that mode and cannot be constructed in analogy to, or *a fortiori* on the basis of, the corresponding generalizations or classifications of the lower mode.

Fourth, no formula for progress from mode to mode, that is, no *strictly genetic* formula in evolution or in development, is possible except by direct observation of the facts of the series which the formulation aims to cover, or by the interpretation of other series which represent the same or parallel modes.

We may now take some given illustrations drawn from

the sciences which show that these axioms, although not explicitly recognized save in part here and there, nevertheless have general application, and that their consistent application would throw light on some of the standing puzzles of the theory of science.[1]

§ 9. *Vital Phenomena and the Theory of Genetic Modes*

As between the purely mechanical or mathematical sciences and that of the next ascending set of phenomena, biology, recent discussion is full of illuminating matter which might be cited in support of these principles.

That the synthesis which is called life is different in some respects from that of chemistry is not only the contention of the vitalists, but also the admission of the adherents of a physico-chemical theory of life. In reply to those who think not only that living matter is a chemical compound, but also that there is nothing to add to this chemical formula — when once it is discovered — in order to attain a final explanation of life, we have only to put to them *the further problem of genesis, as over and above that of analysis* — that is, to ask not only for the analytic formula, the chemical formula, for protoplasm, but also for the laws of reproduction and growth, which always characterize life. The cross-section formula must be supplemented by the longitudinal-section formula. Here we discover the fact that the development is by a series of syntheses, each chemical, but each, so far as we know, producing something new — *a new genetic mode*. If this be denied, then we have to ask the chemist to reproduce the series; and if he

[1] Naturally the illustrations given here are from biology, as that science furnishes the text of the present discourse.

claim that this might be done if he knew how, we ask him to *reproduce the series backwards*. Nothing short of this last form of treatment will do for exact quantitative science. As we found above, no formula which cannot be illustrated by the series of changes stated in a reverse order will fulfil the demands of the shorthand of physical science.

Every chemical process, indeed, whether having only one stage of composition, or whether involving many, has its dissolution series as well as its composition series. The series which the life history of the organism represents is, chemically considered, no doubt a composition series; but when the organism dies, the dissolution series is not at all the reverse of the composition series — a back-tracing of life history. If we say that it is, *i.e.*, that the composition and dissolution series go on together, and that it is always simply a balance in favour of the former — then we are dealing with *two cross-section changes, not with the longitudinal development processes at all.* We ask what it is which constitutes the bond holding these two series together, in what we call the growth or development of an organism as an individual. Either, in short, the character of longitudinal change is present in the composition series, construed as a single set of chemical terms, in which case the dissolution does not reverse it; or the series is a reverse composition series, in which case there is no longitudinal or genetic character about it at all.[1]

What the formula for the longitudinal or strictly genetic

---

[1] This point becomes very much stronger when we cite the racial or evolution series, with the 'immortality' of protoplasm. Think of producing the phenomena of sexual reproduction from mature son to infant father instead of the reverse!

series which represents vital growth and development may turn out to be, no one can tell beforehand simply from the formulas drawn from physics and chemistry, just by reason of this fundamental inability of such formulas to exhaust an irreversible series. The fact that it is irreversible is itself the fact of genetic importance; for it shows that the later terms have some character which the earlier have not.[1]

The second of our axioms, however, must also be true, and it bears directly in the opposite direction — toward the confirmation of the claim of the physico-chemical theory for those phenomena of life to which retrospective and analytic formulas have legitimate application. The data of all science are, as we have seen, subject to this demand. Looked at as an accomplished fact, a life-phenomena is as much a fact subject to the laws of cause and effect and conservation of energy as are the phenomena in any other cases involving physical and chemical constituents. The alternatives are often considered: on the one hand, the exclusive recognition of the categories of regularity and uniformity upon which quantitative science rests, that is, the recognition of the vital processes as physical and chemical phenomena solely, and, on the other hand, the reverse — the introduction into the body of every living cell of a 'somewhat' altogether unamenable to law, and not capable of being recognized by positive science at all. But this antithesis is quite unnecessary; we are not shut up to these alternatives.

As we have seen, the right of physics and chemistry to the universal application of their formulas to their

---

[1] Yet fully admitting the right of quantitative science to show, if it can, that it is reversible.

material, is necessary to the maintenance even of the
genetic point of view as developed in the preceding
section. The new genetic mode is the outcome of the
old. Each has its twofold character; its present organization and its future development. The formulations of
quantitative science are formulations of the organization
— *given the mode*. The mode is the statement of new
form — *liable to organization*. Each, as in Aristotle's
theory, is one aspect of the full truth. This is true also
from the point of view of the rise in the mind of the
distinction upon which analytic science is distinguished
from genetic — that between the retrospective and the
prospective points of view. The very basis of the prospective attitude is found in the formulations which are
retrospective. All the accommodations by which selective thinking proceeds are projected from the platform
of old habitual actions. It is as impossible to construe
the one without the other as to construe the other without the one. To think is at once to recognize both the
analytic and the genetic points of view.

§ 10. *Theories of Life, Mechanical and Vitalistic*

No better illustrations could be wanted of the need of
somehow holding together the two points of view on this
general question of life than the current discussions of
certain critical biological phenomena, such as those of
regeneration. The recent book by Morgan[1] not only lays
before us the data of research, but brings to an issue the
rival theories. We find the advocates of the chemico-
mechanical theory claiming that the data must be con-

[1] *Regeneration*, by T. H. Morgan, 1901.

strued under the law of cause and effect as formulated
in physics and chemistry. Yet they give no adequate
explanation of the remarkable behaviour of the organism
in regenerating its parts. The vitalists, on the other
hand, resort to a view of a highly mystical character,
holding that the organism does what it is its nature to
do, and that no light can be shed upon its behaviour by
the principles of physics and chemistry. An interesting
transition from one of these extremes to the other, in
the same author's views, is to be found in the writings
of Driesch, who works out a theory which attempts to
hold to the adequacy of the formulas of physics and
chemistry in his *Analytische Theorie*, — which, by the
way, outdoes all the metaphysicians for stretches of pure
metaphysics, — and then in later writings goes over
gradually, in the presence of the astonishing revelations
of research, to a frankly vitalistic view. The conclusions
arrived at by Morgan show a somewhat vacillating attempt to do justice to both points of view, at the same
time that a guiding principle whereby they can be rec-
onciled is quite absent. He says: "The fundamental
question turns upon whether the development of a specific form is the outcome of one or more 'forces,' or
whether it is a phenomenon belonging to an entirely
different category from anything known to the chemist
and the physicist. If we state that it is the property
of each kind of living substance to assume under certain
conditions a more or less constant specific form, we
only restate the result without referring the process to
any better known group of phenomena. If we attempt
to go beyond this, and speculate as to the principles
involved, we have very little to guide us. We can, how-

ever, state with some assurance, that at present we cannot see how any known principles of chemistry or of physics can explain the development of a definite form by the organism or by a piece of the organism. Indeed, we may even go farther, and claim that it appears to be a phenomenon entirely beyond the scope of legitimate explanation, just as are many physical and chemical phenomena themselves, even those of the simplest sort. To call this a vitalistic principle is, I think, misleading. We can do nothing more than claim to have discovered something that is present in living things which we cannot explain and perhaps cannot even hope to explain by known physical laws" (p. 255). This seems to concede the main claim of the vitalists. Yet, later on, we find these words: "To prevent misunderstanding, it may be added that while, from the point of view here taken, we cannot hope to explain the behaviour of the organism as the resultant of the substances that we obtain from it by chemical analysis (because the organism is not simply a mixture of these substances), yet we have no reason to suppose that the organism is anything more than the expression of its physical and chemical structure. The vital phenomena are different from the non-vital phenomena only in so far as the structure of the organism is different from the structure of any other group of substances' (pp. 280 f.). This seems to concede the claim of the physicochemical theory except for the reservation regarding structure; and this reservation is most wisely made. For it is just this reservation which, from the point of view of this work, completely neutralizes the claim that an explanation is nothing other than a reduction of a whole to its elements. Such a claim leaves the entire point argued in the earlier pages of this work untouched, *i.e.*, that a real genetic series exhibits *new forms of organization*, new genetic modes, while not violating the laws of the material which is organized.

Consequently, it is quite on the right side to attempt to carry further a theory of the actual method of the organization in the lines of physical explanation. Morgan does this by the suggestion of a series of 'tensions,' made in the last chapter of his book. To be sure, it amounts to little more than suggesting a new term and with it a certain way of looking at the phenomena of development, serving the turn which the fine word 'polarity' also served; yet the approach from the side of physics is justified, so long as the problem of genetic mode — the interpretation of the longitudinal series — is not surreptitiously brought in under that attempt, and smothered under the new term.

What the biologists need to do is to recognize the limitations of one method, and the justification of the other in its own province. In the life processes there seems to be a real genetic series, an irreversible series. Each stage exhibits a new form of organization. After it has happened, it is quite competent to show, by the formulas of chemistry and physics, that the organization is possible and legitimate. Yet it is only by actual observation and description of the facts in the development of the organism, that the progress of the life principle can be made out. The former is quantitative and analytic science; the latter is genetic science.[1]

---

[1] Morgan's somewhat biassed and decidedly inadequate discussion of the natural selection theory of the origin of regeneration seems to show his failure to recognize the need of naturalistic explanations.

§ 11. *Other Applications*

A similar state of things, in another of the most interesting and important spheres of biological discussion, is illustrated by the discussion of natural selection and teleology in an earlier connection. It is there argued that the genetic point of view may be necessary in the consideration of the series of events by which the individual life is accomplished. There may be a form of teleology, or realization of purpose, in the individual's development, and no other sort of explanation than the genetic statement of its modes may be possible. Yet, with it all, when we secure statistical results, we find that they are amenable to statement in a law or curve of distribution, which is the same as if they were due in their origin to mechanical distribution, like the running of shot through a sieve. In other words, the cross-section of the results, after they have taken place, is what physics and chemistry might have produced; but that it arises from the acts that it does, could only be found out by genetic description and investigation. That it would be capable of the teleological construction, or that it is actually brought about by a teleological method, could never be discovered by quantitative science at all.

So also, to cite still another biological case: the new methods of treating biological variations by statistical formulas reach results of value and generality; yet they must rest upon a sufficient number of cases, all at the same level — *all in the same genetic mode* — to justify the quantitative method. The genetic method remains, just the same, the exclusive resort of the historical naturalist, who raises such questions as that of the direction of variations,

their correlations, and in general morphological questions as such.   For all these questions deal with the
development of new genetic modes.   It is, indeed, a valid
criticism of much of the work done by the new methods
that it assumes that the variations which are tabulated
do represent the same modes, or stages of development or
evolution.   If we accept the view that all characters of
the organism are in part epigenetic, — are the outcome
of hereditary impulse plus the bionomic conditions under
which this impulse develops, — then variations themselves
differ at each stage of the individual's development,
differ with age, growth, etc.   In gathering, tabulating,
and treating variations, therefore, only those can be put
together which belong to the same stage; and this is
most difficult to determine.   Suppose we undertake, for
example, to measure the variations in the length of nose
of a species, it will not do to take the noses of individuals at different ages, which represent different stages
of maturity; nor will it do, on the other hand, to take
noses from different environments, where different reactions
have occurred and different amounts of growth in this
direction or that have been possible.   So as to the deduc-
tions which may be made from such measurements for the
theory of evolution, there may be very different formulas
of variation at different evolutionary stages, grades, or
modes.   Influences which are very powerful in effecting
variation in simple organisms, may be largely ineffective
at later periods.[1]

[1] I am not sufficiently versed in their results to judge whether these conditions are sufficiently allowed for; but it seems to be a legitimate demand to
make of the statistician of biological measurements, that all his cases be at the
same stage of development, and that they all occur in common environing
conditions.   Since writing this passage, I have come upon the article by

These illustrations, drawn from biology, may serve the purpose of showing the sort of application the axioms stated above may have. They might be illustrated with great force by questions which involve the relation of biology to psychology, of psychology to ethics, etc. But these topics are too far remote from the main discussions of this work. It may be allowable, however, to point out that the principles stated above as third and fourth are especially *apropos* of certain recent topics of much discussion.

The fourth axiom lays stress upon the actual tracing of each genetic series as it occurs, for itself; but it recognizes the possible sameness of mode in different series which are parallel or, in the sense of an earlier definition, 'concurrent.' The case in biology and also in psychology in which this possibility is realized is that of psychophysical parallelism as worked out and defended in our earlier

---

E. Warren (on 'Variation in the Parthenogenic Generations of Aphis') in Vol. I., Part 2, of *Biometrica* (the journal recently established for the publication of biological measurements), in which he recognizes the requirement, here laid down, that the same conditions of environment should hold for all the cases treated (see especially p. 146 of his article ; see also Weldon's criticism of the 'Mutation theory' in the same journal, I., 3, p. 367). Possibly the other point — that requiring the same stage of development — is also recognized. It would still seem, however, to be almost impossible to fulfil these requirements. At any rate, although we recognize fully the value of the quantitative studies of the new science of 'Biometrics,' and concede that in problems for which the statistical data are adequate it introduces a new era into biology; yet we hold that it illustrates just the point made here — that quantitative science deals with cross-sections, with accomplished organizations, not with transitions and growths as such. The business of the old-school naturalist, who has not the training to do work in 'Biometrics,' is not entirely ruined. And we may express the hope that among the brilliant formulations of 'biometricians' we may not find too many that may be termed *bio-meretricious !* Some such have been produced in fact in the attempt to construct a 'Psychometrics.'

chapters. If parallelism be true, then each mode in one series has theoretically — and many are known to have as matter of fact — a correlated mode at the corresponding level in the other series; and each may, in so far, be used to aid in the interpretation of the other. This holds not only of the parallelism between mind and body, but also of the concurrence between development and evolution.

The third axiom stated above forbids the method of analogy from a lower mode to a higher, either in the solution of a problem of genesis inside a single group or series, or as between one science and another. This, I take it, is the bane of contemporary science other than physical — the carrying over of established formulas, or the analogous application of established principles, often with the question-begging application of the same terms, from one mode of phenomena to another. The theory of evolution is responsible for much of this cheap apology for science — biology used in sociology, physics in psychology, the concept of energy in history, etc. Evolution has been mistaken for reduction, the highest genetic modes being 'explained' in terms of the lowest, and much of the explaining done by 'explaining away' most that is characteristic of the highest. And biological or organic evolution itself is a storehouse of mistaken analogies brought over into the moral sciences.

It is the writer's hope — to close with a personal word — that the series of books, to which this volume belongs, may have done something to show the spuriousness of this sort of science, and to set forth the requirements, at any rate, of what may properly be called genetic investigation.

APPENDICES

APPENDIX A

ORIGINAL STATEMENTS OF ORGANIC SELECTION AND OR-
THOPLASY MADE INDEPENDENTLY BY PROFESSORS H. F.
OSBORN AND C. LLOYD MORGAN, WITH CITATIONS ALSO
FROM PROFESSOR E. B. POULTON

I. Professor H. F. Osborn

['A Mode of Evolution requiring neither Natural Selection nor the Inheritance of Acquired Characters.' (Organic Selection.) *Trans. New York
Academy of Science* (1896), meetings of March and April, 1896, pp. 141–148;
cf. abstract in *Science*, April 3, 1896.]

" Dr. Graf discussed the views of the modern schools of
evolutionists, and adopted the view that the transmission of
acquired characters must be admitted to occur. He cited
several examples which seemed to support this view, and
especially discussed the sucker in leeches as an adaptation to
parasitism and the evolution of the chambered shell in a series
of fossil Cephalopods.

" Professor Osborn remarked in criticism of Dr. Graf's paper
that this statement does not appear to recognize the distinction
between ontogenic[1] and phylogenic variation, or that the adult
form of any organism is an exponent of the stirp, or constitution + the environment. If the environment is normal, the
adult will be normal ; but if the environment (which includes
all the atmospheric, chemical, nutritive, motor, and psychical
circumstances under which the animal is reared) were to change,

---

[1] In this paper 'Ontogenic Variation' is used for what we are now calling
'Modification.' See the citation made below from Professor Osborn's paper
in the *Amer. Naturalist.* — J. M. B.

the adult would change correspondingly; and these changes would be so profound that in many cases it would appear as if the constitution, or stirp, had also changed. Illustrations might be given of changes of the most profound character induced by changes in either of the above factors of environment, and in the case of the motor factor or animal motion the habits of the animal would, in the course of a lifetime, profoundly modify its structure. For example, if the human infant were brought up in the branches of a tree as an arboreal type, instead of as a terrestrial, bi-pedal type, there is little doubt that some of the well-known early adaptations to arboreal habit (such as the turning in of the soles of the feet, and the grasping of the hands) might be retained and cultivated; thus a profoundly different type of man would be produced. Similar changes in the action of environment are constantly in progress in nature, since there is no doubt that the changes of environment and the habits which it so brings about far outstrip all changes in constitution. This fact, which has not been sufficiently emphasized before, offers an explanation of the evidence advanced by Cope and other writers that change in the forms of the skeletons of the vertebrates first appears in ontogeny and subsequently in phylogeny. During the enormously long period of time in which habits induce ontogenic variations, it is possible for natural selection to work very slowly and gradually upon predispositions to useful correlated variations, and thus what are primarily *ontogenic variations* become slowly apparent as *phylogenic variations* or congenital characters of the race. Man, for instance, has been upon the earth perhaps seventy thousand years; natural selection has been slowly operating upon certain of these predispositions, but has not yet eliminated those traces of the human arboreal habits, nor completely adapted the human frame to the upright position. This is as much an expression of habit and ontogenic variation as it is a constitutional character. Very similar views were expressed to the speaker in a conversation recently held with Professor Lloyd Morgan, and it appears as if a similar conclusion had been arrived at independently. Professor

Morgan believed that this explanation could be applied to all cases of adaptive modification, but it is evident that this cannot be so, because the teeth also undergo the same progressively adaptive evolution along determinate lines as the skeleton, and yet it is well known that they do not improve by use, but rather deteriorate. Thus the explanation is not one which satisfies all cases;[1] but it does seem to meet, and to a certain extent undermine, the special cases of evidence of the inheritance of acquired characters, collected by Professor Cope in his well-known papers upon this subject."

['Organic Selection.'[2]   From *Science*, N. S., Vol. VI., pp. 583–587, Oct. 15, 1897.]

" The evidence for definite or determinate variation has always been my chief difficulty with the natural selection theory, and my chief reason for giving a measure of support to the Lamarckian theory. This evidence has steadily accumulated in botanical and zoological as well as paleontological researches, until it has come to a degree of demonstration where it must be reckoned with.[3]

" Quite in another field, that of experimental embryology and zoology, the facts of adaptation to new and untoward circumstances of environment have begun to constitute a distinct and novel series of problems. In many cases they are so

[1] These cases do militate against Lamarckian inheritance, but do not, I think, furnish exceptions to the operation of organic selection; for the deterioration of the teeth by use would only make more necessary the coöperation of muscular and other accommodations, while variations in the teeth were accumulating (cf. the discussion of 'coincident variation' above, Chap. XIV. § 3, and also that of the universal application of the principle above, Chap. III. §§ 1, 5). — J. M. B.

[2] A discussion introduced by Professor Henry F. Osborn and Professor Edward B. Poulton, Detroit Meeting, Amer. Assoc., Aug. 11, 1897.

[3] Cf. Chap. XII. § 1 above, where it is pointed out that Professor Osborn is here possibly using the phrase 'determinate variation' somewhat loosely for 'determinate evolution' — in my opinion a different thing. It is necessary to say this to ·make entirely valid his kind citations from me. Professor Weldon's exposition of the theory of the 'mean' in *Biometrica*, I. 3, may be consulted. — J. M. B.

z

remarkable and so unexplainable that certain German writers, such as Driesch, have taken the ground that they spring from the ultimate constitution of living matter and are incapable of analysis. At the same time it has been recognized that these adaptations are purely individual, transitory, or ontogenic, leaving, for a long time at least, no perceptible influence upon the hereditary constitution of the organism. What may be called the 'traditional' side of these adaptations impressed itself strongly upon Professor James Mark Baldwin in his studies of mental development, also upon Professor Lloyd Morgan in his studies of instinct. The latter, moreover, was one of the first among English selectionists to consider 'determinate variation' as a fixed problem which must be included in any evolution theory. Thus, independently, Professors Baldwin and Morgan and myself put together the facts of individual adaptation with those of determinate variation into an hypothesis which is in some degree new. The first illustration which I used was that of the creation of an 'arboreal man' out of any present terrestrial race by the assumption of an exclusively tree life. This life would be profound in its influences upon each generation producing what would be pronounced by zoologists a distinct speeific type. In course of many thousand years such a type might become hereditary by the slow accumulation of arboreal adaptive and congenital variations.

"Organic selection is the term proposed by Professor Baldwin and adopted by Professor Morgan and myself for this process in nature which is believed to be one of the true causes of definite or determinate variation. The hypothesis is briefly as follows: That ontogenetic adaptation is of a very profound character. It enables animals and plants to survive very critical changes in their environment. Thus all the individuals of a race are similarly modified over such long periods of time that very gradually congenital or phylogenetic variations, which happen to coincide with the ontogenetic adaptive variations, are selected. Thus there would result an apparent but not real transmission of acquired characters.

" This hypothesis, if it has no limitations, brings about a very unexpected harmony between the Lamarckian and Darwinian aspects of evolution, by mutual concessions upon the part of the essential positions of both theories. While it abandons the transmission of acquired characters, it places individual adaptation first, and fortuitous variations second, as Lamarckians have always contended, instead of placing survival conditions by fortuitous variations first and foremost, as selectionists have contended."

[From the *American Naturalist*, November, 1897.]

" On April 13, 1896, I formulated the matter in a paper before the Academy entitled ' A Mode of Evolution requiring neither Natural Selection nor the Inheritance of Acquired Characters,' which has since appeared in *Science*.[1] Professor Baldwin, of Princeton, and Professor Lloyd Morgan, of University College, Bristol, had at the same time independently reached the same hypothesis, and Professor Baldwin has aptly termed it ' Organic Selection.' Both writers have presented valuable critical papers upon it, including in *Science* and *Nature* a complete terminology for the various processes involved. I concur entirely in their proposal to restrict the term ' Variation ' to congenital variation, to substitute the term ' Modification ' for ontogenic variation, and to adopt the term ' Organic Selection ' for the process by which individual adaptation leads and guides evolution, and the term ' Orthoplasy ' for the definite and determinate results.

" The hypothesis, as it appears to myself, is, briefly, that *ontogenic adaptation is of a very profound character; it enables animals and plants to survive very critical changes in their environment. Thus all the individuals of a race are similarly modified over such long periods of time that, very gradually, congenital variations which happen to coincide with the ontogenic adaptive modifications are collected and become phylogenic. Thus there would result an apparent but not real transmission of acquired characters.*

[1] Cited *in extenso* above.

"What appears to be new, therefore, in Organic Selection is, first, the *emphasis* laid upon the almost unlimited powers of individual adaptation; second, the extension of such adaptation without any effect upon heredity for long periods of time; third, that *heredity slowly adapts itself to the needs of a race in a new environment along lines anticipated by individual adaptation, and therefore along definite and determinate lines.*

"Professor Alfred Wallace has recently indorsed this hypothesis in a review of Professor Morgan's work, *Habit and Instinct*, in the March, 1897, number of *Natural Science* in the following language: 'Modification of the individual by the environment, whether in the direction of structure or of habits, is universal and of considerable amount, and it is almost always, under the conditions, a beneficial modification. But every kind of beneficial modification is also being constantly effected through variation and natural selection, so that the beautifully perfect adaptations we see in nature are the result of a double process, being partly congenital, partly acquired. Acquired modifications thus help on congenital change by giving time for the necessary variations in many directions to be selected, and we have here another answer to the supposed difficulty as to the necessity of many coincident variations in order to bring about any effective advance of the organism. In one year favorable variations of one kind are selected and individual modifications in other directions enable them to be utilized; in Professor Lloyd Morgan's words: "Modification *as such* is not inherited, but is the condition under which congenital variations are favored and given time to get a hold on the organism, and are thus enabled by degrees to reach the fully adaptive level." The same result will be produced by Professor Weismann's recent suggestion of "germinal selection," so that *it now appears as if all the theoretical objections to the "adequacy of natural selection" have been theoretically answered.*' (Italics our own.)

"Moreover, in course of discussion of this subject with my friends Professors Lloyd Morgan, Baldwin, and Poulton, a very fundamental difference of opinion becomes apparent; for they agree in believing that the power of plastic modification to new circumstances, or what the Rev. Dr. Henslow has termed 'self-adaptation,' is in itself a result of natural selection. In other words, they hold that natural selection has established in organisms this power of invariable[1] response to new conditions, which, in the vast majority of cases, is essentially adaptive. I disagree with this assumption *in toto*, maintaining that this plastic modification is, *so far as we know*, an inherent power or function of protoplasm. This view, I understand, is also held by Driesch, E. B. Wilson, T. H. Morgan, and probably by many others. The only cases in which self-adaptation may be demonstrated as produced by natural selection are where organisms are restored to an environment which some of their ancestors experienced. We can then imagine that the adaptive response to the old environment is something which has never been lost, as in the well-known reappearance of the pigment in flounders.

"It may be urged against the Morgan, Baldwin, Poulton view that the remarkable powers of self-adaptation, which in many cases are favorable to the survival of the individual, are in many cases decidedly detrimental to the race, as where a maimed or mutilated embryo by regeneration reaches an adult or reproductive stage. It is obvious that reproduction from imperfect individuals would be decidedly detrimental, yet, from the view taken by the above authors, such reproduction would be necessary to secure the power of plastic modification for the race.

"It is certain that, at the present time, one of the surest and most attractive fields of inductive research, leading towards the discovery of the additional factors of evolution, or what I have elsewhere called 'the unknown factor,' is in experimental embryology and experimental zoology. If we could formulate the laws of self-adaptation or plastic modification, we would be decidedly nearer the truth. It appears that Organic Selection is a real process, but it has not yet been demonstrated that the powers of self-adaptation which become hereditary are only accumulated by selection."

[1] Variable (?). — J. M. B.

II. Professor C. Lloyd Morgan

[Extract from *Habit and Instinct* (1896), pp. 312 ff., previously printed by request in *Science*, Nov. 20, 1896, pp. 793 ff., and delivered as one of a series of 'Lowell Lectures' in Boston, early in 1896.]

" In his Romanes lecture, Professor Weismann makes another suggestion which is valuable and may be further developed. He is there dealing with what he terms 'intra-selection,' or that which gives to the individual its plasticity. One of the examples that he adduces is the structure of bone. 'Hermann Meyer,' he says,[1] 'seems to have been the first to call attention to the adaptiveness as regards minute structure in animal tissues, which is most strikingly exhibited in the structure of the spongy substance of the long bones in the higher vertebrates. This substance is arranged on a similar mechanical principle to that of arched structures in general; it is composed of numerous fine bony plates, so arranged as to withstand the greatest amount of tension and pressure, and to give the utmost firmness with a minimum expenditure of material. But the direction, position, and strength of these bony plates are by no means congenital or determined in advance; they depend on circumstances. If the bone is broken and heals out of the straight, the plates of the spongy tissue become rearranged so as to lie in a new direction of greatest tension and pressure; they can thus adapt themselves to changed circumstances.'

" Then, after referring to the explanation by Wilhelm Roux of the cause of these wonderfully fine adaptations, by applying the principle of selection to the parts of the organism in which, it is assumed, there is a struggle for existence among each other, Professor Weismann proceeds to show[2] 'that it is not the particular adaptive structures themselves that are transmitted, but only the quality of the material from which intra-selection forms these structures anew in each individual life. . . . It is not the particular spongy plates which are trans-

[1] Romanes Lecture on *The Effect of External Influences on Development* (1894), pp. 11, 12.
[2] Romanes Lecture, p. 15.

mitted, but a cell mass, that from the germ onward so reacts to tension and pressure that the spongy structure necessarily results.' In other words, it is not the more or less definite congenital adaptation that is handed on through heredity, but an innate plasticity which renders possible adaptive modification in the individual.

" This innate plasticity is undoubtedly of great advantage in race progress. The adapted organism will escape elimination in the life struggle ; and it matters not whether the adaptation be reached through individual modification of the bodily tissues or through racial variation of germinal origin. So long as the adaptation is there, — no matter how it is originated, — that is sufficient to secure survival. Professor Weismann applies this conception to one of those difficulties which have been urged by critics of natural selection. 'Let us take,' he says,[1] 'the well-known instance of the gradual increase in development of the deer's antlers, in consequence of which the head in the course of generations has become more and more heavily loaded. The question has been asked as to how it is possible for the parts of the body which have to support and move this weight to vary simultaneously and harmoniously if there is no such thing as the transmission of the effects of use or disuse, and if the changes have resulted from processes of selection only. This is the question put by Herbert Spencer as to "*coadaptation*," and the answer is to be found in connection with the process of intra-selection. It is by no means necessary that all the parts concerned — skull, muscles, and ligaments of the neck, cervical vertebræ, bones of the four limbs, etc. — should simultaneously adapt themselves *by variation of the germ* to the increase of the size of the antlers ; for in each separate individual the necessary adaptation will be temporarily accomplished by intra-selection,' that is, by individual modification due to the innate plasticity of the parts concerned. ' The improvement of the parts in question,' Professor Weismann urges, ' when so acquired, will certainly not be transmitted, but yet the primary variation is not lost. Thus when an advantageous

[1] Romanes Lecture, pp. 18, 19.

increase in the size of the antlers has taken place, it does not lead to the destruction of the animal in consequence of other parts being unable to suit themselves to it. All parts of the organism are in a certain degree variable (*i.e.* modificable), and capable of being determined by the strength and nature of the influences that affect them, and this capacity to respond conformably to functional stimulus must be regarded as the means which make possible the maintenance of a harmonious coadaptation of parts in the course of the phyletic metamorphosis of a species. . . . As the primary variations in the phyletic metamorphosis occurred little by little, the secondary adaptations would as a rule be able to keep pace with them.'

" So far Professor Weismann. According to his conception, variations of germinal origin occur from time to time. By its innate plasticity the several parts of an organism implicated by their association with the varying part are modified in individual life in such a way that their modifications coöperate with the germinal variation in producing an adaptation of double origin, partly congenital, partly acquired. The organism then waits, so to speak, for a further congenital variation, when a like process of adaptation again occurs ; and thus race progress is effected by a series of successive variational steps, assisted by a series of coöperating individual modifications.

" If now it could be shown that, although on selectionist principles there is no transmission of modifications due to individual plasticity, yet these modifications afford the conditions under which variations of like nature are afforded an opportunity of occurring and of making themselves felt in race progress, a farther step would be taken toward a reconciliation of opposing views. Such, it appears to me, may well be the case.[1]

" To explain the connection which may exist between modifica-

---

[1] In an article entitled 'A New Factor in Evolution,' published in the *American Naturalist* for June and July, 1896, Professor Mark Baldwin has given expression to views of like nature to those which are here developed. And Professor Henry F. Osborn, in a paper read before the New York Academy of Sciences, propounded a somewhat similar theory, but with, he tells me, less stress upon the action of natural selection.

tions of the bodily tissues, due to innate plasticity and variations of germinal origin in similar adaptive directions, we may revert to the pendulum analogy. Assuming that variations do tend to occur in a great number of divergent directions, we may liken each to a pendulum which tends to swing, — nay, which is swinging through a small arc. The organism, so far as variation is concerned, is a complex aggregate of such pendulums. Suppose, then, that it has reached congenital harmony with its environment. The pendulums are all swinging through the small arcs implied by the slight variations which occur even among the offspring of the same parents. No pendulum can materially increase its swing ; for since the organism has reached congenital harmony with its environment, any marked variation will be out of harmony, and the individual in which it occurs will be eliminated. Natural selection then will insure the damping down of the swing of all the pendulums in comparatively narrow limits.

"But now suppose that the environment somewhat rapidly changes. Congenital variations of germinal origin will not be equal to the occasion. The swing of the pendulums concerned cannot be rapidly augmented. Here individual plasticity steps in to save some members of the race from extinction. They adapt themselves to the changed conditions through a modification of the bodily tissues. If no members of the race have sufficient innate plasticity to effect this accommodation, that race will become extinct, as has indeed occurred again and again in the course of geological history. The rigid races have succumbed ; the plastic races have survived. Let us grant, then, that certain organisms accommodate themselves to the new conditions by plastic modifications of the bodily tissues—say by the adaptive strengthening of some bony structure. What is the effect on congenital variations? Whereas all the other pendulums are still damped down by natural selection as before, the oscillation of the pendulum which represents variation in this bony structure is no longer checked. It is free to swing as much as it can. Congenital variations in the same direction as the adaptive modification will be so much to the good of the individual concerned. They will constitute a congenital predisposition to that strengthening of the part which is essential for survival. Variations in the opposite direction, tending to thwart the adaptive modification, will be disadvantageous, and will be eliminated. Thus, if the conditions remain constant for many generations, congenital variation will gradually render hereditary the same strengthening of bone structure that was provisionally attained by plastic modification. The effects are precisely the same as they would be if the modification in question were directly transmitted in a slight but cumulatively increasing degree; they are reached, however, in a manner which involves no such transmission.

"To take a particular case: Let us grant that in the evolution of the horse tribe it was advantageous to this line of vertebrate life that the middle digits of each foot should be largely developed, and the lateral digits reduced in size; and let us grant that this took its rise in adaptive modification through the increased use of the middle digit and the relative disuse of the lateral digits. Variations in these digits are no longer suppressed and eliminated. Any congenital predisposition to increased development of the mid-digit, and decreased size in the lateral digits, will tend to assist the adaptive modification and to supplement its deficiencies. Any congenital predisposition in the contrary direction will tend to thwart the adaptive modification and render it less efficient. The former will let adaptive modification start at a higher level, so to speak, and thus enable it to be carried a step farther. The latter will force it to start at a lower level, and prevent its going so far. If natural selection take place at all, we may well believe that it would do so under such circumstances.[1] And it would work along the lines laid down for it in adaptive modification. Modification would lead; variation follow in its wake. It is not surprising that for long we believed modification to be transmitted as hereditary variation. Such an interpretation of the facts is the simpler and more obvious. But simple and obvious interpretations are not

[1] Professor Weismann's 'germinal selection,' if a *vera causa*, would be a coöperating factor, and assist in producing the requisite variations.

always correct. And if, on closer examination in the light of fuller knowledge, they are found to present grave difficulties, a less simple and less obvious interpretation may claim our provisional acceptance."[1]

'NEW STATEMENT' FROM PROFESSOR LLOYD MORGAN[2]

1. On the Lamarckian hypothesis, racial progress is due to the inheritance of individually acquired modifications of bodily structure, leading to the accommodation of the organism or race to the conditions of its existence.

2. This proposition is divisible into three : (*a*) Individual progress is due to fresh modifications of bodily structure in accommodation to the conditions of life. (*b*) Racial progress is due to the inheritance of such newly acquired modifications. (*c*) The evolution of species is the result of the cumulative series —

$a > b + a' > b' + a'' > b'' + a''' > b'''$, etc., etc., where a, a', a'', a''' are the acquisitions, and b, b', b'', b''' the cumulative inherited results.

3. Anti-Lamarckians do not accept (*b*) and (*c*). But they accept (*a*) in terms of survival. No one denies that individual survival is partially due to fresh modifications of bodily structure in accommodation to the conditions of life.

4. It logically follows from 3 that individual accommodation

---

[1] See also Professor C. Ll. Morgan's later statements in his work *Animal Behaviour* (1900), pp. 37–39, 115.

[2] The above exposition of his position comes to me from Professor Lloyd Morgan after the page-proofs of the body of the book are already passed — in response to my request for annotations on the proofs of Chapter XIV. I have much pleasure in printing it, with Professor Morgan's permission, and regret that I cannot take more direct account of it in the chapter mentioned. It appears to sharpen the definition and also the limitation of the phrase 'coincident variation,' and to set the views of Weismann in a somewhat different relation to organic selection from that which is expressed on pp. 183 ff. above. Whatever the relation may be historically, logically it is certainly close, and the present writer is not at all disposed to be strenuous for an opinion on such a matter.

is a factor in survival which coöperates with adaptation through germinal variation.

5. Weismann, following the lead of Roux, interpreted individual modification in terms of intra-selection. He clearly saw the implication given in 4 above. Speaking of 'the well-known instance of the gradual increase in the development of deer's antlers,' he says (*Romanes Lecture*, 1894, p. 18): 'It is by no means necessary that all the parts concerned should simultaneously adapt themselves *by variation of the germ* to the increase in size of the antlers; for in each separate individual the necessary adaptation [accommodation] will be temporarily accomplished by intra-selection — by the struggle of parts — under the trophic influence of functional stimulus.'

6. So far there is no direct relation between specific modifications and specific variations. Individual accommodation, as a factor in survival, affords time (Weismann, *op. cit.*, p. 19) for the occurrence of *any* variations of an adaptive nature.

7. My own modest contribution to the further elucidation of the subject is the suggestion (1) that where adaptive variation v is similar in direction to individual modification m, the organism has an added chance of survival from the coincidence $m + v$; (2) that where the variation is antagonistic in direction to the modification, there is a diminished chance of survival from the opposition $m - v$; and hence (3) that coincident variations will be fostered while opposing variations will be eliminated.

8. If this be so, many of the facts adduced by Lamarckians may be interpreted in terms of the survival and gradual establishment of coincident variations by natural selection under the favourable environing conditions of somatic modifications.

9. It is clear that there is nothing in this suggestion of a direct relation between specific accommodation and coincident variation which can be antagonistic to the indirect relation indicated above in 6.

10. Correlated and coexistent variations would have the same relations to coincident variations as obtain in other cases of natural selection.[1]

---

[1] Nos. 6, 9, 10 bear upon Chapter XIV. above. — J. M. B.

III. PROFESSOR E. B. POULTON

[From report in *Science*, Oct. 15, 1897, of proceedings of the American Association for the Advancement of Science.]

"Edward B. Poulton, M.A., F.R.S., Hope Professor of Zoology in the University of Oxford, continued the discussion. He began by saying that it must be admitted that the adaptation of the individual to its environment during its own lifetime possesses all the significance attributed to it by Professor Osborn, Professor Baldwin, and Professor Lloyd Morgan. These authorities justly claim that the power of the individual to play a certain part in the struggle for life may constantly give a definite trend and direction to evolution, and that, although the results of a purely individual response to external forces are not hereditary, yet indirectly they may result in the permanent addition of corresponding powers to the species, inasmuch as they may render possible the operation of natural selection in perpetuating and increasing those inherent hereditary variations which go farther in the same direction than the powers which are confined to the individual.

" Professor Osborn's metaphor in opening this discussion puts the matter quite clearly and will be at once accepted by all Darwinians. If the human species were led by fear of enemies or want of food to adopt an arboreal life, all the powers of purely individual adaptation would be at once employed in this direction and would produce considerable individual effects. In fact, the adoption of such a mode of life would at first depend on the existence of such powers. In this way natural selection would be compelled to act along a certain path, and would be given time in which to produce hereditary changes in the direction of fitness for arboreal life. These changes would probably at first be chiefly functional, as Mr. Cunningham has argued (in the Preface to his Translation of Eimer). On these principles we can understand the arboreal kangaroo (Dondrolagus) found in certain islands of the Malay Archipelago, which is apparently but slightly altered from the terres-

trial forms found in Australia. Professor Osborn has alluded
to the arboreal habits said to have been lately acquired by
Australian rabbits ; these and the similar modifications in
habits of West Indian rats are further examples of individual
adaptive modification which may well become the starting-
point (in the sense implied above) of specific variation led
by natural selection in the definite direction of more and
more complete adjustment to the necessities of arboreal life.
Although this conclusion seems to me to be clear and sound,
and the principles involved seem to constitute a substantial
gain in the attempt to understand the motive forces by which
the great progress of organic evolution has been brought about,
I cannot admit that the importance of natural selection is in
any way diminished. I do not believe that these important
principles form any real compromise between the Lamarckian
and Darwinian positions, in the sense of an equal surrender
on either side and the adoption of an intermediate position.
The surrender of the Lamarckian position seems to me complete, while the considerations now advanced only confer added
significance and strength to the Darwinian standpoint.

" I propose to devote the remainder of the time at my disposal
in support of the conclusion that the power of individual adaptation possessed by the organism forms one of the highest
achievements of natural selection, and cannot in any true
sense be considered as its substitute. Professor Baldwin and
Professor Lloyd Morgan thoroughly agree with this conclusion,
and have enforced it in their writings on organic selection.
The contention here urged is that natural selection works upon
the highest organisms in such a way that they have become
modifiable, and that this power of purely individual adaptability
in fact acts as the nurse by whose help the species, as the
above-named authorities. maintain, can live through times in
which the needed inherent variations are not forthcoming, but
in part acts also as a substitute, not indeed for natural selection,
but for the ordinary operation by which the latter produces
change. In this latter case natural selection acts so as to
produce a plastic adaptable individual which can meet any of

the various forces to which it is likely to be exposed by producing the appropriate modification, and this, it is claimed, is in many instances more valuable than the more perfect, but more rigid, adjustment of inherent variations to a fixed set of conditions.

"A good example of the eminent advantages of adaptability in many directions over accurate adjustment in fewer directions is to be found in a comparison between the higher parts of the nervous system in insects and birds. The insect performs its various actions instinctively and perfectly from the first. It is almost incapable of education and of modifying its actions as the result of the observation of the effects of some new danger. It would appear that the introduction of the electric light can only affect the insects which are most attracted to it, by the gradual operation of natural selection. In the clothes-moths which infest our houses, we may see an example of this ; for these insects seem to be comparatively indifferent to light. Birds, on the other hand, have the power of learning from experience, of reasoning from the results of observation. At first terrified by railway trains, they learn that they are not dangerous, and cease to be alarmed ; while the effect of fire-arms results in their increased wariness.

"If this view of individual adaptability as due to natural selection be not accepted, it may be supposed that the individual modifications are due either to the direct action of the external forces or to the tendencies of the organism. But it is impossible to understand how the mechanical operation of such forces as pressure, friction, stress, etc., continued through a lifetime, could evoke useful responses, or why the response should just attain and then be arrested at a level of maximum efficiency. The other supposition, that organisms are so constituted that they *must* react under external stimuli by the production of new, useful characters, or the useful modification of old ones, seems to me to be essentially the same as the old 'innate tendency toward perfection' as the motive cause of evolution — a conception which is not much more satisfactory than special creation itself. The inadequacy of these views is clearly shown when we consider that the external forces which awake response in an organism generally belong to its inorganic (physical or chemical) environment, while the usefulness of the response has relation to its organic environment (enemies, prey, etc.). Thus one set of forces supply the stimuli which evoke a response to another and very different set of forces. We can, therefore, accept neither of the suggestions which have been offered. Useful individual modifications are not directly due to the external forces, and are not due to the inherent constitution of the organism.

" The only remaining hypothesis is that which I have already mentioned, — the view that whenever organisms react adaptively under external forces they do so because of special powers conferred on them by natural selection. This hypothesis will, it seems to me, meet and satisfactorily explain all the facts of the case, whether employed as a preparation or as a substitute for hereditary variations accumulated by natural selection." [1]

[1] In the *Dictionary of Philosophy*, art. ' Plasticity,' the present writer points out that the original mobility or plasticity of living matter is probably different from the more specific plasticity of the developed organisms. — J. M. B.

APPENDIX B

I. F. W. HEADLEY[1]

Influence of the Individual on the Evolution of the Race

" A VARIATION, if it is to forward the process of evolution, must have selection-value in the first individuals in which it appears. A mere rudiment, to be some day, when fully developed, useful to far-off descendants, is not a thing that a clear-headed evolutionist can speak of seriously. The fore-limb of the avian-reptilian ancestor of birds must have been serviceable to him as an oar or a wing or as a compound of the two. It cannot have been a reptile fore-limb spoiled and a mere prophecy of a wing. However imperfect, its usefulness must have been in the present, not in the future. When new circumstances arise there must be, in individuals that are to survive, a fairly complete adaptation ready to hand. The antelopes cannot say to the cheetahs, 'Give us a respite of a hundred generations and we shall be able to race you.' Somehow the antelope has found a way out of the difficulty. Evolutionists have not always been so successful in showing how a species is able to stave off an imminent peril and obtain a respite during which a lucky variation may appear to save it. But now Professor Mark Baldwin[2] and Professor Lloyd Morgan[3] have independently arrived at a theory that makes matters much easier. ' Though there is no transmission of modifications due

---

[1] From *Problems of Evolution* (1901), by F. W. Headley, 1901, Chap. IV., vii, pp. 120 f.

[2] *American Naturalist*, June, July, 1896.   [3] *Habit and Instinct*, p. 315.

to individual plasticity,' writes the latter, 'yet these *modifications* afford the conditions under which *variations*[1] of like nature are afforded an opportunity of occurring and of making themselves felt in race progress.'

" The significance of this principle is clearly seen when it is studied in connection with the family system that prevails among the higher classes of animals, which feed and tend their young and to some extent educate them. Among social species it rises to still greater importance. In the light of this new principle the tending of the young by the parents is not merely a system by which waste is prevented; it is also a system which prevents a species from deviating widely from the line of development that it has begun to follow.

" I shall now try to make clear, mainly by examples, how the principle works. And first I shall try to show its operation when parental affection is not present to bring out its further possibilities. It may be stated thus : *A congenital variation, in itself too minute to affect the question of survival, may gain selection-value through exercise. The variation having thus been saved by exercise, further variations in the same direction may occur.*

" The ancestors of the amphibians lived throughout their lives in water, breathing the oxygen dissolved in it by means of gills. Now individuals in whom a rudimentary lung appeared, a pouch opening from the œsophagus, might develop the breathing capacity of this rudiment by coming frequently to the surface and inhaling air, or by getting out on to the bank either to rest or to escape from enemies. Then there might arise a terrible emergency such as comes to many ' water breathers,' if they live in fresh-water pools ; there might be a drought causing the pools to dry up. At this crisis some individuals are saved by their lungs. They have so far developed their makeshift pouches by exercise that they are able, though not without strain and discomfort, to become exclusively air-breathers, till at length rain comes or they have made their way to another pool from which the water has not evaporated. If there is a succession of such droughts, there will be a further selection

[1] Italics mine (Headley).

of those who have serviceable lungs. Thus individuals tide over a crisis by improving their natural gifts by exercise; without such Lamarckian methods, they would not be equal to the emergency. At the same time, there is a selection of those who can thus improve themselves. When the next drought comes probably further variations in the same direction have arisen, and there would have been an opportunity for this, but for those modifications due to exercise which secured a respite for the species. And thus modifications though not transmitted to the next generation are the prelude to variations similar in tendency to themselves. Before going further, I must say something to justify the above illustration. It is probable that the lung was, in origin, a fully-developed swim-bladder. But a fully-developed swim-bladder may be only rudimentary when regarded as a lung. There was need of exercise to make it serviceable and give it selection-value in this capacity. I have felt justified, therefore, in speaking of it for the sake of simplicity as a rudimentary lung.

"One more instance. I imagine the Wapiti deer, or rather one of his progenitors, — this is the old puzzle set to Neo-Darwinians by Mr. Herbert Spencer, — developing great antlers through the accumulation of congenital variations by Natural Selection. What if the muscles and ligaments of the neck and of all the coöperative machinery did not grow strong through favourable variations during the same period? The answer is plain enough; even without the help of Natural Selection the organism will be able to make shift for a time. Muscles can be strengthened by use during the lifetime of the individual. How much can be done in this way if we begin, say in our teens, and exercise certain muscles regularly for half an hour a day! How great would be the result if we exercised them each day during the whole time that we were on our legs! All day the stag was carrying his antlers, and his muscles were acquiring the strength that was needed. But when the antlers in the course of many generations had grown big, males that were born without specially adapted muscles to carry them would not be likely to be lords of the herd. So that here, too, congenital variations would follow in the wake of accommodations, due to exercise, in the individual.

" The other examples which I take will show how parental affection gave a new importance to this principle.

" First I will consider the process by which birds became bipeds, using their hind-limbs only for walking, and devoting their fore-limbs to flight. Let us assume that they first learned to fly by flapping along the surface of water, flying with their wings and paddling with their feet. When they took to living on land, not only would flight, being unaided by the feet, be more difficult, but they must become bipeds else their wing feathers will suffer. Now walking on the hind legs is by no means an easy feat for a bird till he has been specially adapted for it. What a clumsy creature a penguin is on land! How often he trips and tumbles! But power of running is often indispensable to a bird; many birds in the present day rise from the ground with difficulty, and without ample space cannot rise at all, so that unless they were good on their legs, they would be as helpless as a Boer without his horse. Much less could the primitive bird when he emerged on to the land do without speed of foot, unless like the penguins he was lucky enough to have no land enemies to pursue him. He must, therefore, practise and improve at running, and the result might well be that a small peculiarity of structure would be raised to importance; having a slight gift for running he would become through much practice an adept according to the primitive avian standard. And now comes in the factor of parental training, for we must imagine that having advanced so far in strength, skill, and vitality as to be able to fly, he will not leave his young to fend entirely for themselves. They will have the path of life marked out for them by their parents. They must not return to an aquatic existence, only occasionally landing for rest, at safe spots, but they must be able to stand, walk, run in biped fashion. In fact individuals dictate to their offspring what mode of life they shall follow, choose the environment that is to act upon them, and, each generation making a similar choice, development proceeds cumulatively along certain lines; only variations adapted to the chosen environment are selected, and in a long series of generations the structures and qualities most in demand are brought to a high pitch of excellence. Two more examples will help to make this clearer.

"Imagine the progenitors of the heron taking to fishing in the heron style. As preliminaries they must have some favourable variations; a length of leg beyond the normal, a corresponding length of neck, — this is desirable if not essential, — and also a beak not entirely of the wrong kind. But they do not walk on stilts like their modern descendants, nor have they the other excellencies with which we are familiar. However, by painstaking effort they get over their difficulties and survive in virtue of their piscatorial skill. Moreover, they dictate to their young that they shall be fishermen, and shall fish too in the heron style; no diving is allowed. A propensity to live on carrion is severely discouraged, though a variety of live food, including lizards, insects, and worms is permitted. Among the young some will be failures *qua* herons, will fall short of their parents' almost inadequate development; their neck and legs will suggest anything but fishing in the only style admissible. Nevertheless they will be taken to the water; from the water must come their main food supply. But those that have the heron build, being at the worst not inferior to their parents, will be successes in the line marked out for them; and thus a heron species, afterwards to be dignified as a genus containing many species, will be founded, with long legs, long necks, and ferocious bills.

"One more example may be very briefly given. Let us imagine our own supposed ancestors, tree-climbing animals for long ages, at length taking to walking biped-fashion upon the ground, because the change of habit offered better chances of obtaining food. The new gait would require a whole set of adjustments, for an upright posture is by no means so simple a thing as it seems. It requires certain favourable congenital variations, among others a certain hardness of the soles of the feet, or a tendency to harden under certain conditions. Otherwise lameness would ensue; disease or capture by enemies would follow in due course. Now among the offspring of a pair who succeeded in this new mode of life, some would have feet of the right sort, together with the other characteristics required, and would survive. Others would be ill-adapted for an upright posture and the associated habits. Nevertheless they would have to follow their parents' mode of life. The species, at the time of which we are thinking, has long advanced beyond the stage at which the young are flung upon the world directly they are born. They cannot, therefore, revert to the trees because walking is painful, and, at last, impossible for them. Their parents choose for them their line of life, thus deciding with certain limits the line of development of the species.

"Not only parental care but also the gregarious habit, so common among animals, helps, so to speak, to give the species a continuous policy and so to promote evolution. Here, too, a few examples may make clear what is meant.

"Among a herd of primitive ruminants some individual bulls may have had, where the horns now are, an exceptional thickness of bone and over it a certain epidermal callosity, not sufficient without special treatment to enable them to drive rivals from the field, but sufficient to make them enjoy sparring, so that the parts would get hardened and enlarged during their advance to maturity. Those of inferior natural endowments would improve much less quickly or break down altogether. Thus congenital hardness of head, *increased by practice in butting*, would become a character having selection-value, and bulls that were not richly endowed in this respect would leave no offspring behind them. Among animals living in herds there are special facilities for sexual selection by battle. All the males must fight or efface themselves : there is no standing out. And thus if they are to survive, males must vary in a certain direction, viz., towards hardness of head and weapons for butting. Hence by gradual accumulation will arise horns or branching antlers.

"Many species of birds owe their success in the world to their sociability. Rooks (*Corvus frugilegus*) in their crowded rookeries, or as they fly in large flocks, are able to beat off their enemies. When a party of curlews (*Numenius arquata*) are feeding, it is almost a certainty that one of the number, by means of some sense or other, will become aware of any danger that is approaching and give warning to the rest. And the sociability that thus protects them we cannot regard as entirely instinctive: it is partly habit learned in each by the young from their elders. And thus it comes about that the tradition of the species to some extent decides the course of its evolution by deciding the manner of life that its members are to lead. Those to whom life in a community proves uncongenial probably fall victims to enemies.

" Can this newly discovered principle help to heal the feud between the followers of Weismann (the Neo-Darwinians) and the Lamarckians? If it can, then the Lamarckians must have a singular power of mistaking an utter rout for a compromise. For what the new principle shows is not that acquired characteristics can be transmitted, but that Natural Selection can, without such transmission, do what Lamarckism claimed that it had the exclusive right and power to do. Each generation decides in the main the environment of the next, and insists that it shall live in that environment. Those of the young survive who, with the aid of some training, are able to accommodate themselves to the environment in which they are put. The similarity of environment in each generation leads to selection for similar characteristics: modifications and accommodations in the individual, though not transmitted, are followed by variations and adaptations in the race, the very phenomenon which has always been the Lamarckian's most formidable weapon. This can now be explained on Neo-Darwinian principles, and if you can show that your opponent's theory is not the only nor the best way of accounting for the facts which he himself adduces, you cut the ground from under him. A simile may perhaps make it clear how the principle works. We may look upon a species as a huge herd of animals that are being driven along a road ; the driver being some impulse in themselves. Numerous roads lead off on either side, and it is impossible to say that one is the main road more than another. All these ways lie open, but the elders, by example and persuasion, lead the young into some road or roads swerving at no very great angle from that already followed or into the one that leads straight on. Since the young are not allowed to follow their devious caprices, it is seldom that individuals are found pressing into widely divergent paths. And so the species does not waste itself by vaguely experimenting in new directions. And hence, too, Natural Selection, a policeman who lynches all who don't go the pace or who take a wrong road, works, in a limited field, among the masses that crowd the track that continues the line already followed or others that diverge but slightly from it : among these masses it acts with the utmost stringency ; the laggards are ruthlessly cut off, and evolution goes rapidly on.

" This, I believe, fairly represents the process of evolution in the higher species. But the Lamarckian may fairly enter a demurrer and say : " Low down in the animal scale, the new principle can work but feebly, if at all. There Natural Selection acts directly on the individual from the moment of his birth or the moment of the depositing of the egg. And yet there have been developed forms as high as the newt and the lizard, — an enormous advance from the lowest types. Can Natural Selection have achieved all this ? If not we must find something that will assist it at every stage from the bottom to the top ; not a principle which does not begin to operate till the higher levels have been already attained."

" This objection certainly requires answering.[1] Let us recur to our simile which represents a species as a herd driven along a road from which many roads lead off. If the elders do not guide the young there will be perpetual deviations, most of them ending in wholesale destruction till some guiding tendency develops and is fostered by Natural Selection. This guiding tendency is rigid instinct, and even that does not prevent a slaughter, mainly during infancy, enormously above what takes place among the higher classes of animals. As the crowd

[1] It is claimed by the present writer that this objection is met by the claim that all characters in their development in the individual require some accommodation. This gives organic selection its chance everywhere. — J. M. B.

presses onward, those that, passing all other roads, keep on in a particular direction will at length form a species guided by instincts that seldom swerve. Thus evolution proceeds by Natural Selection, but at the cost of an enormous sacrifice of life, even after instincts come in to reduce it. At the higher level there is intensification of Natural Selection, but the waste of indiscriminate destruction in a great degree comes to an end. Intelligence and plasticity are the order of the day. The monkey is a good representative of the new system: the caterpillar, with his one accomplishment, of the old. Intelligence enables those who have it to make themselves at home where the creature of instinct would perish. They pass their youth in playing and imitating and thus gain a versatility that protects them amid the shocks of circumstance. They have merit of a kind that must make itself felt. Though they have marked tendencies, strong likes and dislikes, yet they have with all a certain saving pliancy and elasticity. And greater pliancy in its component individuals leads, as I have tried to show in this section, to greater adaptability in the species. The result is that among the higher plastic classes of animals evolution proceeds more rapidly.

"But obviously the quickening up of evolution is not all. The individual gains in importance. He improves his powers, is able to face a change of environment that otherwise would have been fatal. He makes an environment for his young in which intelligence can be developed: he chooses the environment which they shall have when out of the nursery, and so decides to some extent what qualities shall be the winning qualities in life. In fact, he is beginning to take the helm and steer the species. Or we may put it in this way: when the individuals of one generation decide the environment in which the next shall grow up, selection ceases to be purely natural: it is in part artificial."

II. Professor Conn [1]

" One of the most recent contributions to the method of evolution has the merit of having been conceived independently by three different naturalists, and recognized from the first as a factor of significance by prominent advocates of both the Neo-Darwinian and Neo-Lamarckian schools. It has been called *organic selection*. The sources from which this idea sprung were quite different, its authors being, one a psychologist, one a paleontologist, and the third a naturalist who has made a special study of instincts. From such different standpoints the arguments that have led to the theory have been somewhat varied. In general it may be said that these naturalists came to this theory because they felt the inadequacy of Natural Selection, as previously understood, to account for all the facts, and because they felt that the Lamarckian factor is at least doubtful, and, even if true, is perhaps not sufficient to meet the demands made upon it. The theory of organic selection is, in a sense, a compromise between the views of the two chief schools. With Neo-Darwinism it abandons the inheritance of acquired characters, but with Neo-Lamarckism it puts the influence of acquired characters foremost in guiding the course of evolution."

Ontogenetic Variations

" In the first place a sharper contrast than ever is drawn between such variations as result from heredity and those which arise from the direct action of the environment upon the individual. This is, of course, simply the difference between congenital and acquired variations, but the latter are now regarded as forming a much larger share in the make-up of an individual than has previously been supposed. The life of an individual may be supposed to begin at the time of the fertilization of the egg. By this time all the hereditary traits that he is to receive are already combined in the egg, *i.e.*, all his congenital characters are within him. But from this moment there begin to act upon

[1] From *The Method of Evolution*, by H. W. Conn, 1900, pp. 303 f.

him the direct influences of the environment, and all subsequently developed variations are acquired rather than congenital. They are frequently called *ontogenetic variations*,[1] which
is a better term than acquired, since all variations must of course
be acquired at some time, and the term ontogenetic indicates
that they are acquired by the individual and not by the germinal
substance. These ontogenetic variations are entirely independent of those which arise in the germ plasm, since they are supposed to affect the body simply and are perhaps not transmitted
by heredity. But such variations have a very great influence
upon the individual. From the very beginning of his life he is
influenced by them, and the characters that he has when adult
are a combination of some that he has received by inheritance
with some which he has developed himself as the result of the
action of the environment upon him. Since these latter characters are the result of the action of the environment, they are
commonly adapted to it. To be sure, as elsewhere pointed out,
we do not understand how environment can act upon the individual in such a way as to produce even acquired adaptive changes
in it. Why a muscle grows with use or diminishes with disuse,
why sensations become more acute when exercised, why changes
in food or climate modify colors, why the shapes of leaves and
the length of the beaks of birds change with climate, we have
not the faintest notion. But such adaptive changes do appear
during the life of the individual. They form the basis of the
Lamarckian theories and are patent in everyday life.

" It is impossible to determine at present to what extent the
characters of an adult are inherited or congenital, and to what
extent they are readily developed by each individual independent
of inheritance. When we remember what extensive changes
can be produced in an organism by changes in its environment,
and remember that the individual from the outset is acted
upon by the environment, it would seem to follow that its adult
characters must in no inconsiderable degree be simply acquired
rather than congenital. But it is difficult or impossible to dis-

[1] Professor Osborn's term, for which in his later writings he has substituted
the term modification. — J. M. B.

tinguish the two classes. In the studies of variations which
have hitherto been made there has been no attempt to distinguish between them. When it is found that the length of the
beaks of birds varies widely with the climate, or that the length
of the wings or legs shows variations on either side of a mean,
it has been assumed that these are innate differences, and therefore, if selected, are matters of heredity. Most of the significance attached to the statistical study of variation mentioned in
an earlier chapter depends upon this assumption. But it is at
least as probable that the variations are simply due to the action
of the environment, habit, or use, and hence purely acquired.[1]
Most of the studies of variation which have been made, up to
the present, have consisted in recording variations, either great
or small, but without attempt to determine to what extent they
are really congenital, and to what extent due to the action of the
environment upon the individual. Considering the great difference in the relation of the two classes to the problem of evolution; it is evident that no very clear results will be reached until
the two types of variation are more carefully separated.

" Be this as it may, it is certain that the environment has a
great influence upon the development of each individual, independent of his inherited characters. It is equally evident
that these acquired characters must change with every change
of condition or habit. If an animal acquires a new food plant
or a new habitat, if he learns a new method of protecting himself, or if a plant starts to grow in a soil different from that in
which it has hitherto lived, these changes will, of course,
produce their effect, and acquired variations will result. Now,
as we have seen, it is difficult to believe that these variations
will so affect the germ plasm as to be transmitted to the next
generation, but it is equally clear that if the next generation should be placed under the same conditions it would
independently develop similar variations, entirely independent
of heredity. So long as the environment remains the same,
each generation will develop, after its birth as an individual,
the same sort of acquired variations. These, appearing regu-

[1] Cf. the positions taken above, pp. 331 ff. — J. M. B.

larly in subsequent generations, would probably be regarded as inherited, although in reality they are only independently acquired by each individual. They would not be a part of the inherited nature, but only the result of the 'nurture' to which each individual is subjected."

Agency of Acquired Variations in Guiding Natural Selection

" The essence of the theory of organic selection is, that these acquired variations will keep the individuals in harmony with their environment, and preserve them under new conditions, until some congenital variation happens to appear of a proper adaptive character. The significance of this conception is perhaps not evident at a glance. It may be made clear by considering, for illustration, the problem of development of habits and organs adapted to each other. It is impossible to believe that an organ develops before the habit of using it, for if it did it would be useless. On the other hand, the habit of using an organ could not arise before the organ makes its appearance. We must thus believe that the organs and the habit of their use appear together, a very difficult or impossible conception for haphazard variation. Now organic selection tries to show that the adoption of a new habit by an animal will result in the development of structures adapted to the habit, but by a principle that does not involve the inheritance of acquired variations. Assuming that some changes in conditions caused certain animals to adopt a new habit, Weismann's theory would force us to believe that some structural changes would follow, from variations in the germ plasm, which would be parallel to the acquired variations developed by the new habit. But when we conceive, as Weismann must, that congenital variations are indefinite and in all directions, it becomes a matter of infinite improbability to suppose that just the right sort of variations will follow such a change in habit at just the right moment. The Lamarckians, finding that habit and structure follow each other so closely, have felt obliged to assume that the one produces the other, while of course the Weismannians must deny such a conclusion.

" If it were not necessary to assume that a congenital variation appropriate to the habit should follow *immediately* when the habit changes, this difficulty would be greatly lessened. It may be admitted that it is so improbable as to be inconceivable that a new habit should be followed immediately by a congenital change in structure appropriate to the new habit, unless there be some inheritance of acquired variation. But it is quite probable, even upon the principle of haphazard variations in all directions, that some such congenital variation might appear in course of time. If the individuals could be kept in their new habit long enough, it would be pretty sure that eventually some congenital variation would appear of an appropriate character. Now the acquired characters will serve to preserve the individual in the new conditions. When an animal adopts a new habit its body begins to change at once, and he soon acquires a development of his muscles and bones adapted to his new habit. He may, indeed, not transmit these characters, and his offspring may be at birth no better off than he was at birth. Each generation acquires these characters for itself so long as the conditions remain the same. But the new characters, even though not congenital, adapt the individual to its new conditions and enable him to live successfully in these conditions. These individuals are therefore able to contend successfully in the struggle for existence, their acquired characters being just as useful to them as they would have been if congenital. This is repeated, generation after generation, similar acquired characters being redeveloped by each generation.

" Remembering then the great numbers of variations that are constantly occurring as the result of modifications of the germ plasm, it is probable, indeed certain, that after a time some congenital variation will appear which will be of direct use to the animals in their new habits. During all this time the majority of variations will appear and as quickly disappear, since, being of no special use, there will be nothing to preserve them, and cross-breeding will soon eliminate them. But when,

perhaps after hundreds of generations, there does appear a congenital variation which aids the animal in its new habit, — an
old habit by this time, — such variations will be selected and
become a part of the inheritance of the race. The individuals
with these congenital variations will, from the outset, have an
advantage over others, since the congenital variations will
enable them to adapt themselves more closely to the conditions
than would purely acquired characters. Thus the acquired
characters keep the individual alive until the proper congenital
variations appear, and the new habit actually determines what
sort of congenital variations shall be preserved, and guides the
process of evolution.

" Perhaps a concrete case may make this somewhat obscure
theory a little clearer. Imagine, for example, that some change
in conditions forced an early monkey-like animal, that lived on
the ground, to escape from its enemies by climbing trees.
This arboreal habit was so useful to him that he continued it
during his life, and his offspring, being from birth kept in the
trees, acquired the same habit. Now it would be sure to follow
that the new method of using their muscles would soon adapt
them more closely to the duty of climbing. Changes in the
development of different parts of the body would inevitably
occur as the direct result of the new environment, and they
would all be acquired characters. The children would develop
the same muscles, tendons, and bones, since they too lived in
the trees and had the same influences acting upon them. Such
acquired characters would enable the animals to live in the trees,
and would thus determine which individuals should survive in
the struggle for existence, for these modified individuals would
clearly have the advantage over those that stayed on the ground,
or did not become properly adapted to arboreal life by acquired
habits. All this would take place without any necessity for
a congenital variation or the inheritance of any character which
especially adapted the monkey for life in the trees.

" But, in the monkeys thus preserved, congenital variations
would be ever appearing in all directions. It would be sure to
follow that after a time there might be some congenital variation

that affected the shape of the hands and feet. These would not be produced as the result of the use of the organs or as acquired variations, but simply from variations in the germ plasm. There might be thousands of other variations in other parts of the body in the meantime. The miscellaneous variations, however, would not persist. But as soon as variations appeared which affected the shape of the hands and the feet, the fact that the animal had continued to climb trees would make these variations of value, and therefore subject to natural selection. Selection would follow, and thus in time the monkeys might be expected to inherit hands and feet well adapted for climbing. The acquired variations, in such a case, had nothing to do with *producing* the changes directly, but they did shield the animal from destruction until congenital variations appeared. Acquired variations have determined that the individuals shall live in trees, and this life has determined what congenital variations will be preserved. Indirectly, therefore, the acquired variations guide evolution.

" This factor would also aid in explaining the origin of co-ordinated structures, which have been always a puzzle to natural selection. How, for example, can we imagine that chance congenital variations shall *at the same time* cause an increase in the size of the deer's horns and in the strength of his neck and shoulders ? Either without the other could not exist. But we can imagine that some congenital variation increased the size of the antlers, and then clearly enough acquired characters would of necessity increase the size of the neck and shoulder muscles, thus enabling the animal to carry the large antlers. This might continue many generations. Eventually another series of variations of a congenital character might affect these muscles. These would be at once selected, if they enable the animal to carry its antlers more easily, and thus in time neck and antlers would be coördinated to each other. The animal by acquired characters adapts itself to its conditions and waits until a proper congenital character appears. A combination of characters to make a coördinated system of organs is thus made possible, in a manner that natural selection alone is unable to account for."

Consciousness a Factor

" This conception of the action of selection evidently makes consciousness a factor in evolution. It has always been claimed by the Lamarckian school that consciousness aids in the process of descent. It has sometimes been supposed that by this claim is meant that by conscious efforts an animal can modify its structure ; but such a conception has certainly not been held by scientists in recent years. Consciousness may, however, lead to the use of organs or to the adoption of the new habits, and, if the view we are now considering be sound, such use of organs, or such habits, leads to the development of acquired characters which enable the individual to live in new conditions more successfully, until after a time congenital variations take their place. Consciousness thus becomes an indirect factor in evolution. Indeed, the attempt is sometimes made to extend this principle of consciousness to all organic life, and to find even among the lower plants something which corresponds to it. Such an expansion of consciousness is, however, too crude and unintelligible to take its place in our general conception of nature. But, if organic evolution be a factor [fact?], consciousness becomes a force of considerable importance among higher animals. Moreover, this is just where there appears to be the greatest need for some aid to Natural Selection. As we have already seen, there is strong evidence for the inheritance of acquired characters among plants, so strong indeed that some botanists insist that it is a matter of demonstration that such characters are inherited. Among animals, however, there is little evidence for such inheritance and apparently a growing disinclination to believe in it. Thus it is seen that the factor of consciousness would come into play just where acquired characters become of most doubtful value. Among plants, because of the wide distribution of the germ plasm through the body, there is less difficulty in accepting the inheritance of acquired characters, and here consciousness is not needed. Among animals, where the inheritance of acquired characters is more doubtful, to say the least, this factor of consciousness takes its place."

2 B

Organic Selection and Natural Selection

"It has been said that organic selection is a sort of compromise between Weismannism and Lamarckism. It can, however, hardly be called a compromise. It abandons entirely the Lamarckian position of the inheritance of acquired characters, and that such agencies as use and disuse have any direct influence in producing variations which modify the offspring by inheritance. The only Lamarckian feature that is left is, that the environment, through the acquired characters it produces, does have an important influence in guiding evolution. Such a position is, however, perfectly in accordance with Weismannism, as is shown by the fact that organic selection is endorsed by Weismann.[1] At the same time, there is no doubt that it quite materially alters the earlier notions of Natural Selection and presents that theory in quite a different aspect. For it is plain that with this idea the guiding force in evolution is no longer simply the natural selection of minute, haphazard variations, as Darwin supposed, but a combined action of the indirect influence of acquired variations and the selection of haphazard congenital variations. It has long been felt that the theory of evolution by the selection of mere haphazard variations presents great difficulties, and, if it were possible to find some more distinctively guiding force the gravest difficulties of Natural Selection would disappear. It is for this reason that the Lamarckians insist upon acquired variations as a guiding force, and others claim that variations occur along definite lines. This new factor of organic selection tries to show that acquired variations, although not directly inherited, do furnish such a guiding force, since they preserve the life of the individual by adapting him to his new conditions, until a time, after many generations perhaps, when some congenital variations of a proper character appear.

"If this factor of organic selection is admitted as a force, we must ask how wide is its application. Is it a force like Natural Selection, that will apply everywhere, or is it confined, as are the

---

[1] Professor Conn does not say where. — J. M. B.

effects of use and disuse, to certain organisms? In answer to this it is apparent that its influence will be more extended than the action of use and disuse, and more extended than the limits of consciousness. Wherever acquired variations occur, organic selection will apply. Wherever environment, either food, climate, or conscious action, produces direct modifications of the body of animal or plant, these acquired variations will aid in preserving the individual until the proper congenital variations appear. Organic selection would therefore seem to apply wherever the environment produces a direct adaptive variation in the body of the individual."[1]

"Organic selection must undoubtedly be regarded as a factor in the evolution of species. This is granted on all sides. In the study of the history of man it becomes of extreme significance, but this subject cannot be considered in this work."

[1] Cf. the note on p. 360, above. — J. M. B.

APPENDIX C

RECENT BIOLOGY[1]

1. *Année biologique: Comptes rendus annuels des travaux de biologie générale.* YVES DELAGE. Première année, 1895. Paris, Reinwald. 1897. pp. xlv + 732.
2. *Essays.* G. J. ROMANES. Edited by C. LL. MORGAN. London and New York, Longmans, Green & Co. 1897. pp. 253.
3. *Darwin and after Darwin, III. Isolation and Physiological Selection.* G. J. ROMANES. Edited by C. LL. MORGAN. Chicago, Open Court Pub. Co. 1897. pp. viii + 178.

I.

IN this handsome volume (1) Professor Delage begins the annual issue of a summary of biological progress ; a work which was well begun in his earlier volume on *Heredity*, etc. In the preface to this volume we read: " To those who have read the volume on *Heredity and the Great Problems of Biology* this new annual will not cause surprise. It is the natural sequel to that work. . . . The earlier work may be considered as a first volume, serving the purpose of setting the questions, defining the problems, tracing the outlines, establishing the categories, and resuming the results, up to 1895, from which date this periodical takes up all the topics and carries them on from year to year." The reader of the *Grands Problemes* will have, therefore, a fair idea of the divisions, headings, etc., of this new publication. The way in which the general purpose of the editor and his contributors is carried out in this first volume calls for

[1] From *The Psychological Review*, Vol. V., No. 2, March, 1898.

much admiration. Not only will it be of great value to biologists, but students in neighbouring departments, especially in psychology, will find it a reliable and readable introduction to the newer biological problems in their latest phases. One feature strikes the present writer as peculiarly good — albeit exceedingly difficult — *i.e.*, the attempt of the editor to gather up in a few pages a statement of the advance made during the year under each great heading, thus giving a résumé of each of the successive résumés of literature made by the contributors. Such a 'skimming-off of the cream' could only be done by a master, and must in any case involve some personal equation ; but Professor Delage has shown in his earlier work the sort of grasp on the entire subject — both as to information and as to judgment — which such an undertaking demands.

The allowance of space to psychology, under the head of 'mental functions,' is adequate and just. It is sincerely to be hoped that the editor will not take the advice of certain reviewers and restrict this department in future issues. Not only is this section of value to psychologists, as bringing their work into organic connection with biological results, but even more to biologists, who are thus informed of the light which psychology, especially in its genetic and evolutionary phases, is coming to throw on some of the standing problems of biology. This is seen in the volume before us in the statement of recent advances in the questions of instinct, individual adaptation, and determinate evolution.

On the whole, therefore, we may count this publication as a distinct addition to the apparatus of the natural sciences, and extend congratulations to its learned editor and his collaborators.[1]

The two posthumous works of Romanes (2 and 3) are valuable additions to our legacy from that acute mind. The book of essays is less valuable than the other, seeing that it is a collection of papers published at various dates, which do not in all cases represent the latest and most matured opinions of the author. They all have biographical value, however, — meaning

[1] The *Année Biologique* maintains its high excellence from year to year, the fifth volume, 1899–1900, having now appeared (1901).

mental biography, of course, — especially those of a more practical character, which bring out human points of view. In the other work, we have the systematic exposition of the theory of Physiological Selection, which is possibly Romanes' most original and interesting single contribution to natural history.

This theory has two main features; features which should be taken separately, I think, and which only lose in force and serve to introduce confusion when brought under a single point of view, as Romanes does. The real novelty of physiological selection consists in the hypothesis that congenital variations toward infertility might lead to relative segregation in a group of animals living together, and the development of the groups thus segregated away from one another in divergent lines. No one who appreciates the problem of inter-specific infertility can, I think, fail to see the force of this hypothesis, nor fail to agree with Romanes — quite apart from the evidence of fact — in the hypothesis that specific differences may be secondary to sexual variations, rather than the reverse, as Darwin supposed. I cannot help thinking, however, that Romanes places too much confidence in the so-called 'principle' of Weismann (amixia) and Delbœuf, that any slight average difference between different groups must develop itself. That would seem to depend upon circumstances; and at any rate it is purely hypothetical. Romanes weakens his case by making it a sort of corner-stone to his structure; for whatever the causes be of the subsequent divergent evolution — say Wallace's pure utility view — the original segregation by physiological selection would lose none of its value, if it be true, especially in cases of absolute infertility. The value of physiological selection as producing divergent species would seem to rest, in cases of relative or partial infertility, largely upon the sort of variations which were correlated[1] with the infertility — a point which the theory of Reproductive Selection of Professor Karl Pearson covers.

[1] As to whether partial infertility alone, without any regular correlations, would produce divergent results, seems very doubtful, except as it tended to result (by accumulation of variations) in absolute infertility. This latter result Romanes himself supposes.

What the two have in common is the postulate of infertility, Romanes assuming its segregation value, and so finding it available to produce divergent, or as he calls it ' polytypic,' evolution.

The other point of which Romanes makes so much — and, I think, unfortunately — is that in which he agrees with the Rev. Mr. Gulick, the writer who first proposed and has elaborately expounded — but under different terms — the principle of physiological selection. Both of these authors, Romanes later so far as one can gather, formulated the general principle of ' Isolation '; meaning by it — to gather the matter up briefly — any sort of relative control of pairing. If, for any reason, males *A* to *L* can pair with females *a* to *l*, but cannot pair with females *m* to *z*, these males are then ' isolated ' from the latter females. Under this ' principle,' on the author's showing, everything ' in heaven above and on earth beneath ' can be brought. Natural selection is only a case of isolation; so is the migration of Wagner, and the geographical separation of Weismann, and physiological selection from infertility, and artificial, and indeed sexual selection. He says : " Equalled only in its importance by the two basal principles of heredity and variation, this principle constitutes the third pillar of a tripod on which is reared the whole superstructure of organic evolution " (p. 2). With all the laboured proof of this proposition, it suffices to say that it is true, because self-evident; and at the same time, in the present state of biological science, well-nigh worthless. For the very concept of heredity through sexual reproduction presupposes it. All heredity *in particular* involves the ' isolation' of the two parents temporarily for the purposes of the act of mating. We might even go so far as to announce a great ' principle of negative isolation ' (!), *i.e.*, that by artificial selection, or any sort of human regulation, the upper limit to the birth-rate in any species may be set by the isolation of the male from more than one mate. Surely it adds nothing to natural selection to call it also isolation, explaining that it depends upon the elimination of some individuals and the consequent isolation of those not banished to the shades; nor does it add anything to the other sorts of selection, now historic both as facts and as having names, to call them 'isolation.' All this seems to the present writer to furnish evidence of the tendency of Romanes, shown also strikingly in his later writing on the inheritance of acquired characters, to lay too much value on logical disquisition.[1]

In thus dwelling on the striking features of physiological selection, as Romanes and Gulick have developed it, I by no means mean to lead the reader to think that this important theory is done justice to; on the contrary, the book will be found, from many points of view, to build up a claim for this hypothesis as representing a real factor in evolution,—especially in divergent evolution,—which writers who refuse to recognize it, as Mr. Wallace, will have great difficulty in disposing of. And this the more when it is taken in connection with the evidence which Professor Pearson gives to show that 'Reproductive Selection' (on the basis of relative infertility) is actually at work.

For example, among a certain class in a community, a high relative death-rate among women of narrow hips may serve to establish a correlation between maximum effective fertility (in Pearson's sense) and broad hips; while in another class in the same community the same maximum fertility may perhaps be established by intentional regulation of size of family with better medical attendance, without any reference to size of hips at all. Here there would be a tendency to divergent evolution in the matter of hip conformation, due simply to 'isolation' by a social barrier. Romanes' hypothesis calls for the same result where the barrier is the physical one of some degree of gross infertility between the two groups. I put forward this social instance because, among other reasons, while it is one of the few forms

---

[1] As to the minor utility of showing that there is such a wider though negative category under which certain of these natural processes may be viewed — that, no one, I suppose, would dispute ; but when it comes to considering it a great discovery, and requiring biologists to adopt a new terminology with a view to recognizing it, it would seem to be going too far. Nor does this suggest any disparagement of the fresh and new considerations advanced, especially, in Mr. Gulick's very notable papers. A similar classification of certain of the special 'factors' under the general head of 'isolation' is made by F. W. Hutton in *Natural Science*, October, 1897.

of 'isolation' — by a social barrier — which were not already recognized and named, yet it is one of the forms which Romanes and Gulick did not recognize nor name. It is also interesting as showing a type of cases in which groups *living together* (that is, not geographically separated), and at first quite fertile *inter se*, might acquire infertility, as a consequence of other morphological changes, thus illustrating Darwin's view, but under Romanes' conditions. I have called this choosing a mate under social limitations 'personal selection,'[1] but, like all the other 'selections,' it might be scheduled under 'isolation,' of course.

It is interesting, also, to note that Darwin recognized several forms of isolation (see Romanes' quotation, p. 108, note) besides geographical separation; and among them two forms which involve physiological selection, *i.e.*, 'breeding at slightly different seasons,' and 'individuals preferring to pair together' (sexual selection). The latter is a case of physiological selection, if only we make the highly probable assumption that the 'mental preference' for certain mates carries with it maximum fertility with those mates.[2]

[1] In the work, *Social and Ethical Interpretations*, Sect. 40. See also the table of forms of 'Selection' given above in this work, Chap. XII. § 2.

[2] This is a correlation which I have never seen suggested anywhere; yet if it should be true, Mr. Wallace would have to admit physiological selection as a sort of organic counterpart of his selective association by recognition marks. Without such a correlation, sexual preference would seem to lose much of its biological significance. It might get some support from the fact that the coyness of the female, which, on the hypothesis of Groos (*Play of Animals;* see the review of Professor Groos' book following), plays an essential part in sexual selection, demands increased strength and persistence in the male's impulses. It might be made a matter of experiment to determine whether highly coloured, grand-mannered birds are either absolutely or relatively very fertile ; or it might be observed whether sexual-criminals (in whom the impulse on the mental side may be considered strong) have unusually large families, or progeny later in life than others — both, however, very complex problems involving other factors.

II.

Die Spiele der Thiere. By KARL GROOS, Professor of Philosophy in the University of Giessen. Jena, Gustav Fischer. 1896. pp. xvi + 359. (*The Play of Animals*, Eng. trans. by Eliz. L. Baldwin. Appletons, 1898.)[1]

In this volume Professor Groos makes a contribution to three distinct but cognate departments of inquiry — philosophical biology, animal psychology, and the genetic study of art. Those who have followed the beginnings of inquiry into the nature and functions of play in the animal world and in children will see at once how much light is to be expected from a thoroughgoing examination of all the facts and observations recorded in the literature of animal life. This sort of examination Professor Groos makes with great care and thoroughness, and the result is a book which, in my opinion, is destined to have wide influence in all these departments of inquiry.

I cannot take space for a detailed report of Professor Groos' positions. It may be well, therefore, before speaking of certain conclusions which are to me of special interest, to give a résumé of the contents of the book by chapters. Chapter I. is an examination of Mr. Spencer's 'surplus energy' theory of Play; the result of which is, it seems, to put this theory permanently out of court. The author's main contention is that play, so far from being 'by-play,' if I may so speak, is a matter of serious business to the creature. Play is a veritable instinct, true to the canons of instinctive action. This view is expanded in Chapter II., where we find a fine treatment in detail of such interesting topics as imitation in its relation to play, the inheritance of acquired characters *apropos* of the rise of instincts, the place and function of intelligence in the origin of these primary animal activities. This chapter, dealing with the biological theory of play, is correlated with Chapter V., later on in the book, in which the 'Psychology of Animal Play' is

---

[1] From *Science*, Feb. 26, 1897. Portions of this notice were incorporated in the present writer's preface to the English translation.

treated. Together they furnish the philosophical and theoretical basis of the book, as the chapters in between furnish the detailed data of fact. I shall return to the biological matter below. Chapters III. and IV. go into the actual ' Plays of Animals' with a wealth of detail, richness of literary information, and soundness of critical interpretation which are most heartily to be commended. Indeed, the fact that a pioneer book on this subject is, at the same time, one of such unusual value, both as science and as theory, should be a matter of congratulation to workers in biology and in psychology. The collected cases, the classification of animal plays, as well as the setting of interpretation in which Professor Groos has placed them — all are likely to remain, I think, as a piece of work of excellent quality in a new but most important field of inquiry.

As to the plays which animals indulge in, Professor Groos classifies them as follows : ' Experimenting,' ' Plays of Movement,' ' Play-Hunting ' (' with real living booty,' ' with play living booty,' ' with inanimate play booty '), ' Play-fighting ' (' teasing, scuffling among young animals,' ' play-fighting among adult animals '), so-called ' Building Art,' ' Nursing ' plays, ' Imitation ' plays, ' Curiosity,' ' Pairing ' plays, ' Courting by Means of Play of Movements,' ' Courting by the Exhibition of Colours and Forms,' ' Courting by Noises and Tones,' ' Coquetry on the Part of the Female.'

With this general and inadequate notice of the divisions and scope of the book, I may throw together in a few sentences the main theoretical positions to which the author's study brings him. He holds play to be an instinct[1] developed by natural selection (for he does not accept the inheritance of acquired characters), and to be on a level exactly with the other instincts which are developed for their utility. It is very near, in its origin and function, to the instinct of imitation, but yet they are distinct (a word more below on the relation between play and imitation). Its utility is, in the main, twofold : first,

[1] Modified in *The Play of Man* in a way which makes the word 'impulse' a better designation, in the author's maturer view. This substitution of terms may be made throughout this review.

it enables the young animal to exercise himself beforehand in the strenuous and necessary functions of its life and so to be ready for their onset; and second, it enables the animal by a general instinct to do many things in a playful way, and so to learn for itself much that would otherwise have to be inherited in the form of special instincts; this puts a premium on intelligence, which thus comes to replace instinct (pp. 65 f.). Either of these utilities, Professor Groos thinks, would insure and justify the play instinct; so important are they that he suggests that the real meaning of infancy is that there may be time for play.[1]

It is especially in connection with this latter function of play that the instinct to imitate comes in to aid it. Imitation is a real instinct, but it is not always playful; play is a real instinct, but it is not always imitative. Professor Groos does not suggest, I think, closer relations between these two instincts. There is likely, however, to be a great deal of imitation in play, since the occasion on which a particular play instinct develops is often that which also develops the imitative tendency as well, *i.e.*, the actual sight or hearing of the acts and sounds of other animals. Moreover, the acquisition of a muscular or vocal action through imitation makes it possible to repeat the same action afterwards in play.

It is only a step, therefore, to find that imitation, as an instinct, has to have ascribed to it, in a measure, the same race utility as play — that of going before the intelligence and preparing the way for it, by rendering a great number of specialized instincts unnecessary. It is interesting to contrast this view with that which the present writer has recently developed in the pages of *Science* (see Chap. V., above), *i.e.*, the view that imitation supplements inadequate congenital variations in the direction of an instinct, and so, by keeping the creature alive, sets the trend of further variations in the same direction until the instinct is fully organized and congenital. If both

---

[1] "Die Thiere spielen nicht weil sie jung sind, sondern sie haben eine Jugend, weil sie spielen müssen" (p. 68). Other capital utilities which might be added are (1) the exercise of the intelligence itself and (2) direct *social* utility as such.

these views be true, as there seems reason to believe, then imitation holds a remarkable position in relation to intelligence and instinct. It stands midway between them and aids them both. In some functions it keeps the performance going, and so allows of its perfection as an instinct; in others it puts a stress on intelligence, and so allows the instinct to fall away if it have no independent utility in addition to that served by intelligence.[1] In other words, it is through imitation that instincts both arise and decay — that is, some instincts are furthered and some suppressed, by imitation. And all this is accomplished with no appeal to the inheritance of acquired characters, Professor Groos agreeing with Weismann that the operation of natural selection as generally recognized is sufficient.

The difficulty which I see to this conception of play as a pure instinct is that which is sometimes urged also against considering imitation an instinct, *i.e.*, that it has no definite motor coördinations, but has all the variety which the different play forms show. If the definite congenital plays are considered each for itself, then we have a great many instincts, instead of a general play instinct. But that will not do, for it is one of Professor Groos' main contentions, in the chapter on the psychology of animal plays, that they have a common general character which distinguishes them from other specialized instinctive actions. They are distinguished as play actions, not simply as actions. This difficulty really touches the kernel of the matter, and serves to raise the question of the relation of imitation to play; for imitation presents exactly the same conditions — a general instinct to imitate, which is not exhausted in the particular actions which are performed by the imitation.

---

[1] In a private communication Professor Groos suggests that the two views may well be held to supplement each other. The case is very much like that of early intelligence, in the form of association; where it fully accomplishes the utility also subserved by an instinct, it tends to supersede the instinct; otherwise, it tends to the development of the instinct (Groos, p. 64). (See p. 140 above and cf. the same writer's *Play of Man*, translated by the same hand, in which the principal suggestions of this notice have been taken account of by Professor Groos.)

I shall remark on the solution of it below, in speaking of Professor Groos' psychology of play. It will be interesting to see how he treats this problem in his promised work on the *Spiele der Menschen;*[1] for the imitative element is very marked in children's plays.

Other points of great interest in this biological part are the emphasis which Groos finds it necessary to put on 'tradition,' instruction, imitation, etc., in young animals, even in enabling them to come into possession of their natural instincts; in this the book tends in the same direction as the new volume of Professor C. Lloyd Morgan. Again, there is a remarkably acute discussion of Darwin's Sexual Selection, which the author finally accepts in a modified form by saying that the female's selection is not necessarily conscious, but that she has an inherited susceptibility to certain stimulating colours, movements, etc., in the male. It is not so much intelligence on her part as increased irritability in the presence of certain visual and other stimulations.[2] Over against the charms of the male he sets the reserve or reluctance (*Sprödigkeit*) of the female, which has to be overcome, and which is an important check and regulator at the mating time. Again, the imperfect character of most instincts is emphasized, and the interaction with imitation and intelligence. He finds a basis for the inverse ratio between intelligence and instinct in an animal's equipment on natural selection principles, *i.e.*, the more intelligence develops the less does natural selection bear on special instincts, and so they become broken up.

Finally, I should like to suggest that a possible category of 'Social Plays' might be added to Groos' classification — plays in which the utility of the play instinct seems to have reference to social life as such. Perhaps in such a category it might be possible to place certain of the animals' performances which

---

[1] See the note above which indicates that Professor Groos, in the *Play of Man*, considers play an 'impulse,' taking on different forms.

[2] 'Sexual' is thus referred back to 'natural' selection (p. 274), although the direct results of such preferential mating would still seem to give very 'determinate' direction to evolution under natural selection. (Cf. *Science*, Nov. 13, 1896, p. 726; see Chap. XI. § 2, above.)

seem a little strained under the other heads — for example, those performances in which the social function of *communication* is exercised early in life. A good deal might be said also in question of the author's treatment of 'Curiosity' (*Neugier*). He makes curiosity a function of the attention, and finds the restless activity of the attention a play function, which brings the animal into possession of the details of knowledge before they are pressed in upon him by harsh experience. My criticism would be that attention does not fulfil the requirements of the author's psychological theory of play, as indicated below.

Turning now to the interesting question of the psychological theory, we find it developed, as it would have to be, in a much more theoretical way. The play consciousness is fundamentally a form of 'conscious self-illusion' (pp. 311 ff.) — *bewusste Selbsttäuschung*. It is just the difference between play activity and strenuous activity that the animal knows, in the former case, that the situation is not real, and still allows it to pass, submitting to a pleasant sense of illusion. It is only fair to say, however, that Herr Groos admits that in certain definite instinctive forms of play this criterion does not hold; it would be difficult to assume any consciousness of self-illusion in the fixed courting and pairing plays of birds, for example. The same is seen in the very intense reality which a child's game takes on sometimes for an hour at a time. Indeed, the author distinguishes four stages in the transition from instincts in which the conscious illusion is absent, to the forms of play to which we can apply the phrase 'Play activity' in its true sense, *i.e.*, that of *Scheinthätigkeit* (pp. 298 f.). The only way to reconcile these positions that I see is to hold that there are two different kinds of play — that which is not psychological at all, *i.e.*, does not show the psychological criterion at all, and that which is psychological as *Scheinthätigkeit*. Herr Groos does distinguish between 'objective' and 'subjective' *Scheinthätigkeit* (p. 312). The biological criterion of definite instinctive character might be invoked in the former class, and the psychological criterion in the other. And we would then have a situation which is exemplified in many other functions of animal and human life — functions which are both biological and instinctive, and also psychological and intelligent, as sympathy, fear, bashfulness. Then, of course, the further question comes up as to which of these forms is primary, again the old question as to whether intelligence arose out of reflexes or the reverse.

I think some light falls on this time-honoured question from the statement of it in connection with this new question of play, and especially when we remember Herr Groos' theory of the function of imitation and the extension of his view suggested above. If imitation stands midway between instinct and intelligence, both furthering the growth of instinct, and also leading to its decay in the presence of intelligence, then we might hold something like this : In proportion as an action loses its consciously imitative and volitional character, to that degree it loses its *Schein* character, and becomes real in consciousness and instinctive in performance (and this applies to the cases in which imitation has itself become habitual and instinctive) ; and on the contrary, in proportion as an instinctive action is modified and adapted through imitation and intelligence, to that degree it becomes capable of assuming the *Schein* character and is indulged in as conscious play. I cannot enlarge upon this here, but it seems to square with a good many of the facts, both those which Groos cites as showing that imitation opens the way for the decay of instinct with the growth of intelligence, and those which Morgan and I have cited as showing that imitation keeps congenital variations alive and so allows them to accumulate into instincts. And I think it so far confirms the view that imitation is a sort of meeting-point of race habit, represented by instinct, and race accommodation, represented by intelligence — just the double function which imitation serves also in the development of the individual (cf. my volume on *Mental Development, in loc.*).

Going into the analysis of the play psychosis, Herr Groos finds several sources of pleasure to the animal in it (pp. 203 ff.) — pleasure of satisfying an instinct, pleasure of movement and energetic action, but, most of all, ' pleasure in being a cause.' This last, together with the ' pleasure in experimenting,' which characterizes many play activities, is urged with great insistence.

Even the imitative function is said to produce the joy of 'victory over obstacles.' Yet, here again, the author is compelled to draw the distinction between the play which is psychological enough to have a represented object, and the instinctive sort in which the pleasure is only that of the instinct's own performance. The pleasure of overcoming friction of movement, also, is very doubtful, since in any but the instinctive games which are cited (Chap. I.) to prove that the animal is not using up surplus energy (seeing that he plays after he is tired) — in other games we stop playing when the friction and inertia of the muscles become conscious as fatigue. Much more, however, is to be said for the pleasure of rivalry, or of overcoming an opponent, in the higher types of play; but Herr Groos scarcely does this justice.

Returning to the element of illusion in play, we find two ingredients in it (pp. 313 ff.) — a division of consciousness (*Spaltung des Bewusstseins*), *i.e.*, a division between the activity treated as real and the sense that it is unreal. There is considerable oscillation between these two poles. This ability to treat representations as realities is, according to Herr Groos, the essential of all imagination. In play it is akin to the division of consciousness found in certain pathological cases of double personality. It is a sort of hypnotization by the stream of representations, but with the sense that it is all an illusion and may be pierced through by a return to reality at any moment. This seems to me a true and valuable characterization of the play consciousness (it is taken from K. Lange), but Professor Groos' extension of it to all imagination does not seem to hold. In his criticisms of others (as the present writer) he fails to honour the current distinction between 'fancy' and 'constructive imagination.' In fancy we do yield ourselves up to a play of images, but in the imagination of scientific thinking or of artistic creation are not both the goal and the process strenuous enough? This, indeed, leads Professor Groos to a view of art which allies it closely with the play function, but to that I return below.

The second element in the play or '*Schein*' consciousness is the feeling of freedom (*Freiheitsgefühl*; pp. 331 f.). In play there is a sense of 'don't-have-to,' so to speak, which is contrasted

2 C

both with the necessity of sense and with the imperative of thought and conscience. This idea seems to be part of Schiller's theory of play. So Groos thinks the general feeling of freedom holds in consciousness only while there is a play of motives to which the agent may put an end at any moment — a sense of ' don't-have-to ' in the life of choice. This sense of freedom keeps the *Schein* consciousness pure and prevents our confusing the play content with the possible real contents of life. This is very interesting and suggestive. The sense of freedom is certainly prominent in play. Whether it should be identified with the sense of control which has been used by some writers as a criterion (both in a negative and in a positive sense) of the belief in realities already experienced, or again with the freedom with which choice is pregnant, is more questionable. Without caring to make a criticism of Professor Groos' position, I may yet point out the distinction already made above between the two sorts of imagination, one of which has the ' don't-have-to ' feeling and the other of which does not. So also in our choices there are those which are free with a ' don't-have-to ' freedom, but there are choices — and these are the momentous ones, the ones to which freedom that men value attaches — which are strenuous and real in the extreme. Indeed, it seems paradoxical to liken the moral life, with its sense of freedom, to a ' game of play,' and to allow the hard-pressed sailor on the ethical sea to rest on his oars behind a screen of *Schein* and plead, ' I shan't play.' Seriously, this is something like the result, and it comes out again in the author's extremely interesting sections on art, of which I may speak in conclusion.[1]

Those who have read Professor Groos' former stimulating book, *Einleitung in die Æsthetik*, will anticipate the connection which he finds between play and art.[2] The art consciousness is

---

[1] In the later volume, *The Play of Man*, Professor Groos so modifies his definition of play as to make the only criterion what I have called its ' auto-telic ' character, as having its own end (*Selbstzweck*), being performed simply for itself with no further end.

[2] The reader may now consult another later publication by Professor Groos, *Der æsthetische Genuss* (1902).

a consciousness of *Schein;* it is also a play consciousness, inasmuch as it is the work of imagination — both the creative and the appreciative art consciousness — and the meaning of imagination here is that it takes *Schein* for reality. The 'self-conscious illusion' of the play consciousness is felt in extreme form in the theatre, and the pleasure of it is felt even when we play with painful situations, as in tragedy. In art the desire to make an impression on others shows the 'pleasure of being a cause.' This intent to work on others is a necessary ingredient in the art impulse (pp. 312 f.). Groos differs from K. Lange, who holds a similar view of the necessary division of consciousness between reality and *Schein* in the æsthetic psychosis, in that Lange thinks there must be a continual oscillation between the two poles of the divided consciousness, while Groos thinks there is rather a settling down in the state of illusion, as in an artist's preoccupation with his creations, a novelist with his characters, and a child with her doll (pp. 323). In art the other great motive of play, 'experimenting,' is also prominent, and is even more fundamental from a genetic point of view; of that a word below.

Here, again, the question left in my mind is this : whether the play motive is really the same as the art motive. Do we not really distinguish between the drama (to take the case most favourable to the theory) as amusement and the drama as art? And does the dramatist who is really an artist write to bring on self-illusion in the spectator by presenting to him a *Schein* scene ? Possibly, art theorists would divide here ; the realists taking more stock in *Schein*, since realistic art is more nearly exhausted by imitation. This sort of illusion undoubtedly gives pleasure, and it is undoubtedly part of art pleasure. Yet there does seem to be, in a work of fine art, a strenuous outreach toward truth, which is additional — both in the production and also in the enjoyment — to the instrument of appearance used by the artist. It may be that we should distinguish between truth which comes to us didactically and truth which comes artistically, and make the method of the latter, and that alone, the source of æsthetic impression. In any case the theory of Groos, which has its roots in the views of Lange and v. Hartmann, is extremely interesting and suggestive, especially as contrasted with the recent psychological theory of Mr. H. R. Marshall. In the present theory, the 'self-exhibition' of which Mr. Marshall makes so much, enters as the need of impressing others with the play illusion. As to the hedonic element and its ground, however, the two theories are in sharp contrast, and that of Groos seems to me, on the whole, more adequate. In the wealth of literary reference in his book, Mr. Marshall pays singularly little attention to the authors from whom Groos draws, and none to the earlier work of Professor Groos himself, but treats the play theory only in the form of Mr. Spencer's surplus energy construction. To Groos' theory, musical art would present difficulties and so would lower sensuous æsthetic effects generally.

Genetically art rests upon play, according to Herr Groos, in that the three great motives of art production, 'Self-exhibition' (*Selbstdarstellung*), 'Imitation,' and 'Decoration' (*Ausschmückung*), are found in the three great classes of animal plays, respectively, 'Courting,' 'Imitation,' and 'Building Art' (*Baukünste*, seen in birds' nest-building, etc.). On the strength of this, Groos finds both æsthetic appreciation and impulse in the animals, and all rests upon the original 'experimenting' impulse. Of this, however, Professor Groos does not give a satisfactory account. Experimenting is a necessary part of effective learning by 'imitation,' I think, and the use made of it in the selection of movements may be its original use.

On the whole, Professor Groos' book is both a pioneer work and one of great permanent value. It contains a good index and a full list of the literary sources.

III.

Habit and Instinct. By C. Lloyd Morgan, F.G.S. London and New York, Edward Arnold. 1896. pp. 351.[1]

Professor Morgan's *Habit and Instinct* adds another to his series of works, now three in number, dealing with comparative psychology. The reader is impressed anew with the prime

[1] From *The Nation*, May 13, 1897.

quality which he has learned to expect in this author's writing —
great lucidity, secured at once by a simple style, long reflection,
and a certain persistence in making his point tell. Combined
with this is a balance and caution which invites the reader's
confidence, and leaves the impression that the writer, even in
the theoretical parts of his subject, can always be trusted. At
the same time we find that the work goes over many of the
same topics as the earlier books, repeats some of the same
instances, even repeats itself more than is necessary, and while
the net gain is great, — the book is one of the most important
in the recent literature of the problem of instinct, — yet both
the observations and the discussions could have been put into
much less space, for half the price. The volume will tend in
some degree to supersede the one on *Animal Life and Intelligence*, since the author has now reached points of view on the
most important subjects, such as the relation of instinct and
intelligence, the inheritance of acquired characters, imitation,
etc., which render it impossible for workers to quote the earlier
work as representing Professor Morgan's maturer views.[1]

As to the essential teachings of the present book, we have
space to give only their most important bearings in connection
with recent discussion. Among recent publications on this side
of the water Professor Morgan makes use of the observations
of Professor Wesley Mills of Montreal on the instincts and
habits of young animals, and the experiments and conclusions
of the present writer reported in the work on *Mental Development in the Child, etc.* It will be remembered that Professor
Morgan, in a course of Lowell lectures in Boston in 1896,
dwelt on the results of detailed experiments carried out by
him with young birds, artificially hatched and reared under
constant observation. The early chapters in *Habit and
Instinct* contain these experiments carried still further. The
substantial results are in agreement with those of Mills, and
go to show that many of the actions of young fowls, which

---

[1] Professor Morgan has now issued an entirely new work, *Animal
Behaviour* (1900), which, as he tells us, grew out of the attempt to revise
the volume *Animal Life and Intelligence*.

have been considered quite instinctive, — as the experiments of Spalding and others seem to show, — are really a mixture of congenital tendency and acquired habit. Some of these activities are of vital importance, such as drinking, fleeing from constant enemies (as the hawk), appreciating and acting upon exact spatial relationships, etc. Such results, found also in the examination of trustworthy reports of animals, as those of Hudson in the *Naturalist in La Plata*, lead Professor Morgan to his most important conclusions. Briefly stated, they are somewhat as follows : —

First, this imperfection of instinct, even in things vital to the organism, emphasizes the intelligent and imitative learning processes of young animals. These learning processes keep them alive by supplementing their congenital activities and structural capacities. This conclusion gives new importance to the psychological processes. Second, the question arises as to the sort of things which young animals learn and how they can learn them. Upon this, again, observations throw light. The fact appears that there are certain relatively constant functions and activities handed down from generation to generation in animal families and communities, as has been theoretically insisted on by Wallace, and recently confirmed by the observations of Hudson under the term ' tradition,' and by the present writer, who calls the individual's learning of tradition ' social heredity.' And, third, the question of the method of organic evolution has some light shed upon it, in Professor Morgan's opinion, by the relation between these learning processes of the animals and natural selection. Professor Morgan here develops (Chap. XIV.) a suggestion which has also been put forth by Professor H. F. Osborn, and independently reached by the present writer, as Morgan points out, namely, that by learning intelligently and imitatively to do things which are essential, certain animals are screened from the operation of natural selection, and so hand on their capacities to future generations, while the race accumulates further congenital variations in the same directions (what Morgan calls ' coincident variations '). Thus evolution takes the direction marked out in the first instance by the individual's learning.[1] All these writers agree that this suggestion neutralizes in great measure the current arguments for the inheritance of acquired characters, since, if evolution is directed in any case in the channels of the acquired characters in the way suggested, it becomes unnecessary to suppose, in the absence of evidence in favour of it, that the same characters are also directly inherited. It may be noted that among others Mr. A. R. Wallace, in a recent review of *Habit and Instinct* in the journal *Natural Science*, welcomes this suggestion.

Possibly our readers will be most interested in certain positions regarding "Human Evolution" which Professor Morgan reaches (in Chap. XV.) on the basis of the observations and conclusions already briefly set forth. He seems well justified in drawing them in view of the foundation laid in his other chapters. His main contention is that, even in the animal world, the method of learning by the individual — *i.e.*, imitation, association, profiting by experiences of pleasure and pain — is essentially different, and the progress which is secured through it is essentially different, from natural selection and the progress secured through it. In the former, consciousness becomes 'efficient,' at least in a sense. It is not clear to us just how much this means from a philosophical point of view — this 'efficiency' of consciousness — in the mind of Professor Morgan; but it is yet clear that in the case of man, where social transmission comes to replace physical heredity as the means of handing on the mass of tradition and race acquisition, consciousness, whatever it is able to do, has the field largely to itself. In human evolution, therefore, we are not under the law of natural selection alone, operating upon fortuitous variations. We are rather under the law of conscious selection accumulating its stores through social and intelligent handing down. Natural selection weeds out the worst on a large scale; conscious selection picks out the best individuals, the best actions, arrangements, beliefs, etc. This is the way the author and

---

[1] It is from this chapter in Principal Morgan's book that the passage cited above, Appendix A, is taken. — J. M. B.

some of those whom he quotes would reconstruct the relation of biology to social evolution; and the position seems to be fruitful enough.

Readers at all versed in recent biological discussion may remember the sort of fatalistic results which the new Neo-Darwinian theories of human evolution were supposed to bring. If the discipline and the dissipation of parents have little or no effect upon their children, we are asked, where is the place of social reform and the motive to individual training? The answer to this comes through the line of teachings brought together in this book. The individual is not born with a physical heritage increased by his father's acts, but into a social heritage which takes its character from the set of conditions which the father also lived in and contributed to. We all make these conditions better or worse, and we all profit by them for better or for worse, in a new and truer sense. The individual is redeemed from the capricious and accidental effects of single lives lived for good or ill, but he inherits socially the larger influences which make the social environment what it is, and which represent a continuous social movement.

We cannot dwell upon the special question which Professor Morgan discusses with his usual clearness and force — such as the relation of instinct to acquired habit, the function of sexual selection, the details of the specific habits of mammals and birds. These discussions may, however, well be brought to the attention of biologists and psychologists. In conclusion, we may notice emphatically the contrast between this book and the work of those recent writers who deal with the same large questions of heredity, degeneration, race-progress, etc., having only scented biology from afar, and having learned their anthropology from Lombroso and Nordau.

INDEX [1]

Accommodation, 45, 85, 94 f., 111, 151.
Adaptive movements, 54.
Agenetic science, 300.
Analogies, genetic, 43.
Analogous organs, 176.
Aristotle, 295, 327.
Artificial selection, 166, 175.
Ataxic variation, 162.
Attention (as selecting function), 252 ff.;
    (variations in), 256.
Automaton theory, 129.
Autotaxic variation, 162.

Bailey, 229.
Bain, A., 23, 88, 124, 249, 259.
Bather, F. A., 198 f.
Bernheim, 93.
Binet, 92.
Biological and psychological, 4.
Biometrics, 333.
Bionomic, 9.
Bosanquet, B., 268.
Bumpus, 152.
Bunge, 92.

Categories (natural history of), 281 ff.
Cattell, J. M., 154 ff.
Causation (category of), 288 f.
Characters (congenital and acquired), 34.
Circular reaction, 110, 123.
Co-adaptation, 62.
Cockerell, 172.
Coincident variation, 38, 150 f., 196 ff.
Competition (economic), 221.
Concurrence (intergenetic), 11, 41, 106,
    189 ff., Appendix B.
Conn, 116, 125, 156, 170, 182 f.
Conscious selection, 167.
Consciousness (in evolution), 50.
Continuity (equal), 11.
Convergence, 176.
Cope, E. D., 50 ff., 63, 77, 80 ff., 121 ff., 198.

Correlation, 24; (correlated variations),
    196 ff.
Cunningham, 191, 203.

Dallinger, 215.
Darwin, 43, 154 ff., 179, 197, 211, 213, 214.
Darwinism, 50 ff., 135 ff.
Davenport, 33, 95 f., 164, 184.
Defrance, 230.
Delage, 36, 165 f., 183, 209, 230.
Design, 277 f.
Determinate (evolution), 34 ff., 135 ff.;
    (variation), 160 ff.
Development and evolution, 3.
De Vries, 33.
Discontinuous variation, 182.
Divergent evolution, 176.
Dohrn, 227.
Driesch, 183, 328.
Dualism, 129.
Duplicated functions, 72.
Dynamogenesis, 54, 86 f. (see also under
    Functional Selection and *Accommoda-
    tion*).

Eimer, 9, 139, 142, 183.
Emotion, 27.
Epigenesis, 50.
Excess discharge, 88.

Fitness (of thoughts), 258 ff.
Force, concept of, 6.
Fortuitous variation, 231.
Functional selection, 63, 109, 166.

Galton, 58, 146, 211.
Genetic (analogies), 43; (modes, theory
    of), 300 ff.; (science), 308 ff.
Germinal selection, 166.
Groos, K., 29, 32, 140, 155 ff., 174 ff., 207,
    220.
Gulick, Rev. Mr., 172, 176.

[1] The Appendices A, B, C are not indexed.

Index